Graphics Hardware 2004

Eurographics Symposium Proceedings

Grenoble, France, August 29–30, 2004

General Co-Chairs

Michael Doggett (ATI, USA)
Anselmo Lastra (University of North Carolina, USA)

Program Co-Chairs

Mark Harris (NVIDIA, USA)
David Luebke (University of Virginia, USA)

Papers Co-Chairs

Tomas Akenine-Möller (Lund University, Sweden)
Michael McCool (University of Waterloo, Canada)

Proceedings Production Editors

Dieter Fellner (TU Braunschweig, Germany)
Stephen Spencer (The University of Washington, USA)

Sponsored by EUROGRAPHICS Association in cooperation with ACM SIGGRAPH

Werner Hansmann, Werner Purgathofer, François Sillion
Series Editors

Published by the Eurographics Association
PO Box 16, CH-1288, Aire-la-Ville, Switzerland

Printed in Germany

Cover design by Stefanie Behnke and Stephen Spencer

ISBN 3-905673-15-0
ISSN 1727-3471

The electronic version of the proceedings is available from the Eurographics Digital Library at
<http://diglib.eg.org/>.

Table of Contents

Table of Contents

Computation

Preface

These proceedings contain the papers from the 19th Eurographics/SIGGRAPH Graphics Hardware workshop which was held August 29 and 30 in Grenoble, France.

The 14 papers presented were selected from 43 submissions, a record number of submissions for the workshop. The continuing growth and increasing quality of submissions to the workshop made the selection process difficult. The final set contains an excellent selection of work covering the most relevant areas of graphics hardware today. We would also like to thank Stefanie Behnke and Dieter Fellner for helping us with the Eurographics online reviewing system.

Presentations from industry on the latest developments were also presented in the Hot3D track and published separately.

We would like to thank the many people who have made this workshop possible, including the program committee members and external reviewers for their help in reviewing the papers. We would also like to thank Stephen Spencer for his help preparing the proceedings.

Many thanks to our generous sponsors: ATI, HP, Microsoft Research, NVIDIA, S3 Graphics. We wish all participants a stimulating workshop and hope they enjoy the opportunity to create and renew their connections within the graphics hardware community.

Michael Doggett
Anselmo Lastra
General Chairs

Mark Harris
David Luebke
Program Chairs

Michael McCool
Tomas Akenine-Möller
Papers Chairs

Michael Meißner
Publicity Chair

Sylvain Lefebvre
Local Arrangements Chair

Arie Kaufman
Steve Molnar
Wolfgang Straßer
Steering Committee

Sponsors

Graphics Hardware (2004)
T. Akenine-Möller, M. McCool (Editors)

A Quadrilateral Rendering Primitive

Kai Hormann Marco Tarini

Visual Computing Lab, ISTI / CNR, Pisa

Abstract

The only surface primitives that are supported by common graphics hardware are triangles and more complex shapes have to be triangulated before being sent to the rasterizer. Even quadrilaterals, which are frequently used in many applications, are rendered as a pair of triangles after splitting them along either diagonal. This creates an undesirable C^1*-discontinuity that is visible in the shading or texture signal. We propose a new method that overcomes this drawback and is designed to be implemented in hardware as a new rasterizer. It processes a potentially non-planar quadrilateral directly without any splitting and interpolates attributes smoothly inside the quadrilateral. This interpolation is based on a recent generalization of barycentric coordinates that we adapted to handle perspective correction and situations in which a quadrilateral is partially behind the point of view.*

Categories and Subject Descriptors (according to ACM CCS): I.3.1 [Computer Graphics]: Hardware Architecture—*graphics processors*; I.3.3 [Computer Graphics]: Picture/Image Generation—*display algorithms*; I.3.7 [Computer Graphics]: Three-Dimensional Graphics and Realism—*color, shading, shadowing, and texture*

1. Introduction

Quadrilateral meshes are a natural and popular choice for *modelling* many classes of 3D surfaces. For example, they are useful for approximating Bézier patches and to describe rotational surfaces and rectangular height fields. In contrast, current GPUs do not support *rendering* of quadrilaterals directly and whenever a quadrilateral (or *quad*) is displayed, one of several options has to be chosen (see Figure 1).

The most frequently used approach is the **diagonal split** along one of the two diagonals which decomposes the quad into two triangles [SA03]. As a consequence, any attribute interpolation inside the quad will be only C^0-continuous at the chosen diagonal. This C^1-discontinuity is undesirable because it is well visible if, for example, shading is computed from interpolated normals or texture mapping is applied.

A less common technique is to perform **two-fold linear interpolation** with a standard scan-line rasterizer. Attributes are linearly interpolated first along the projected edges and then across each horizontal span from edge to edge [SA03]. Unfortunately, this also creates C^1-discontinuities, namely along the horizontal screen lines that pass through the quad's vertices. These discontinuities are more disturbing because they vary with the viewing direction and even if the camera is just rotated around the z-axis [Duf79].

An effective way to avoid discontinuities is to utilize the **projective map** that maps the quad to the planar unit square and to compute the attributes for a point inside the quad by bilinear interpolation over this standard domain [Duf79]. Although this method interpolates smoothly inside the quad it has several drawbacks: it gives a false impression of depth; it does not interpolate attributes linearly along the edges, thus creating C^0-discontinuities between two neighbouring quads; and it requires the projected quad to be convex.

Another option is to refine the quad into $n \times n$ small quads so as to reduce the artefacts caused by the diagonal split. The positions and attributes of the new vertices can be computed by **bilinear interpolation** or any other subdivision scheme either on-the-fly or by preprocessing. For sufficiently large n, this gives visually smooth results, but since many more vertices and triangles must undergo the vertex processing and triangle setup phase, it can cause a severe performance downgrade, especially for geometry-limited applications.

In this paper we present a new way to render quadrilaterals as atomic primitives that gives results which are often visually similar to the ones obtained by bilinear interpolation and in any case superior to the diagonal split. Moreover, it inherits most of the properties that we are used to from rendering triangles and blends well with the design of current graphics systems.

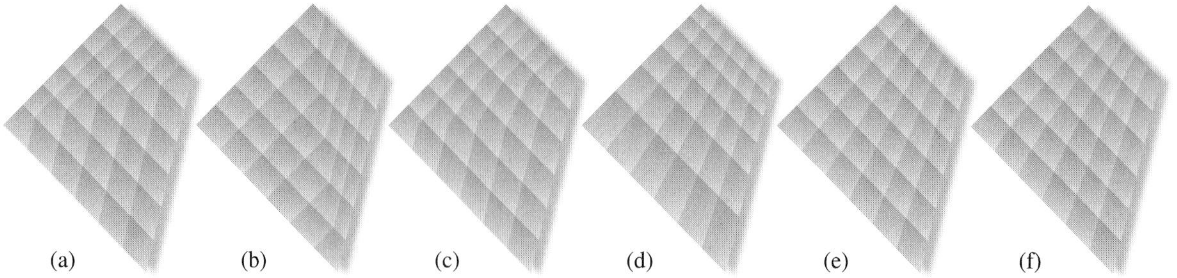

Figure 1: *Texture mapping a regular pattern with different methods onto a flat quad that is seen from above. From left to right: the two diagonal splits (a,b), two-fold linear interpolation (c), projective map (d), bilinear interpolation (e), our method (f).*

Our approach is based on recent theoretical and technological advances. On the one hand, we utilize and extend Floater's *mean value coordinates* [Flo03] which generalize the concept of barycentric coordinates and allow to interpolate attributes smoothly over the interior of a planar quad and linearly along its edges (Section 3). On the other hand, modern GPUs can execute an increased number of complex operations per second. Admittedly, this computational power has mostly been spent on per-vertex and per-fragment processing so far, but it could also be used to implement our method efficiently as a new rasterizer (Section 4). This quad rasterizer would provide a new "tube" for the graphics pipeline between the vertex and the fragment shader that could be used as an alternative to the existing rasterizers for triangles, lines, and points.

2. Main Idea

Our algorithm renders a three-dimensional, potentially non-planar quad $Q = [V_0, V_1, V_2, V_3]$ in the same three steps that are commonly used for triangles:

1. **Vertex shader.** Project the 3D vertices V_i to screen and perform any other per-vertex operation.
2. **Rasterizer.** Create a fragment for every pixel inside the projected 2D quad and use generalized barycentric coordinates to appropriately interpolate the vertex attributes.
3. **Fragment shader.** Process the produced fragments with their interpolated attributes.

The first step is identical to the one that is currently used for other primitives (triangles, lines, and points). A 3D vertex V_i with world coordinates (X_i, Y_i, Z_i) is projected to screen by first multiplying its *homogeneous coordinate vector* with the 4×4 matrix $\mathbf{M}$ that combines the necessary transformations (model-view, perspective, and window-to-viewport),

$$(\tilde{x}_i, \tilde{y}_i, \tilde{z}_i, w_i) = (X_i, Y_i, Z_i, 1)\,\mathbf{M}. \tag{1}$$

The result is then normalized to obtain the *screen position* $v_i = (x_i, y_i) = (\tilde{x}_i/w_i, \tilde{y}_i/w_i)$ and the *depth value* $z_i = \tilde{z}_i/w_i$. In addition, all the *vertex attributes* that shall be interpolated inside the quad (normals, texture coordinates, etc.) are set up per vertex in this step, possibly by a vertex program.

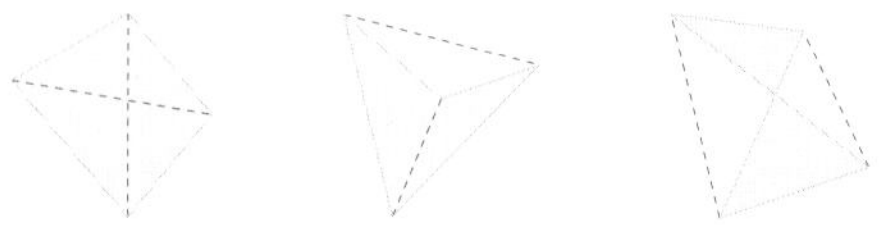

Figure 2: *A quad in front of the camera projects to a convex, concave, or self-intersecting shape in screen space. Dashed lines indicate the quad's diagonals.*

Step two is the core part of our method and although it is similar in spirit, it differs considerably from the triangle analogue. After the four vertices of Q have been projected, we consider the 2D *screen quad* $q = [v_0, v_1, v_2, v_3]$ that can have any of the shapes shown in Figure 2. Finding the interior screen pixels of q and determining their orientation (front- or back-facing) is rather simple and can be done, for example, with a general polygon scan-converter [FvDFH90] or using *edge functions* (Section 4.1). In contrast, interpolating vertex attributes is more delicate and has not been solved satisfactorily, so far.

But why is this interpolation so easy in the case of triangles and yet so hard for quads? The reason is that the underlying concept of barycentric coordinates is trivial for triangles, but it has only recently been generalized to arbitrary convex polygons (see [FHK04] and the references therein). We discovered that the *mean value coordinates* [Flo03], which are one possible generalization, naturally extend to concave and self-intersecting quads (Section 3). We further discuss how they can be adapted to handle perspective correction (Section 3.1) and *exterior quads* (Section 3.2).

The third step proceeds exactly as in current hardware implementations and treats fragments (compute shading, lookup texture, etc.) either with the standard fixed functions or with a user-defined fragment program.

3. Barycentric Coordinates for Quadrilaterals

Given an n-sided planar polygon with vertices $v_0, \ldots, v_{n-1}$, the basic concept of barycentric coordinates is to have for any point v a set of coordinates $\lambda_i(v)$ with the two properties

$$\sum_{i=0}^{n-1} \lambda_i(v) v_i = v, \quad (2)$$

$$\sum_{i=0}^{n-1} \lambda_i(v) = 1, \quad (3)$$

and to use them for interpolating attributes a_i that are associated with the vertices v_i straightforwardly by

$$a(v) = \sum_{i=0}^{n-1} \lambda_i(v) a_i. \quad (4)$$

It is well-known that the barycentric coordinates for a triangle are uniquely defined by the linear functions

$$\lambda_i(v) = A(v, v_{i+1}, v_{i+2}) / A(v_0, v_1, v_2), \quad (5)$$

where $A(u_0, u_1, u_2)$ denotes the *signed* area of the triangle $[u_0, u_1, u_2]$. Apart from the fact that the attribute interpolation (4) can efficiently be computed for these coordinates, their most important properties are:

- **Lagrange property.** Since $\lambda_i(v_j)$ equals 1 if $i = j$ and 0 otherwise, it follows that vertex attributes are reproduced at the vertices, i.e. $a(v_j) = a_j$.
- **Edge linearity.** The λ_i are linear along the edges of the triangle, so that the attribute interpolation is C^0-continuous across the common edge of two neighbouring triangles.
- **Positivity.** The interpolated attribute $a(v)$ of a point v that lies *inside* the triangle is a *convex combination* of the a_i because in this (and only this) case all $\lambda_i(v)$ are positive.

In order to define barycentric coordinates for arbitrary polygons it is often easier to first construct a set of *homogeneous coordinates* $\mu_i(v)$ that satisfy

$$\sum_{i=0}^{n-1} \mu_i(v)(v_i - v) = 0 \quad (6)$$

and then derive the barycentric coordinates by normalization

$$\lambda_i(v) = \mu_i(v) \Big/ \sum_{j=0}^{n-1} \mu_j(v). \quad (7)$$

These $\lambda_i(v)$ obviously fulfil Equations (2) and (3) and a general construction of homogeneous coordinates $\mu_i(v)$ has been discussed in [FHK04] in the case of *convex* polygons. The critical part about using their construction for our purposes is the extension to *concave* or *self-intersecting* quads. For most

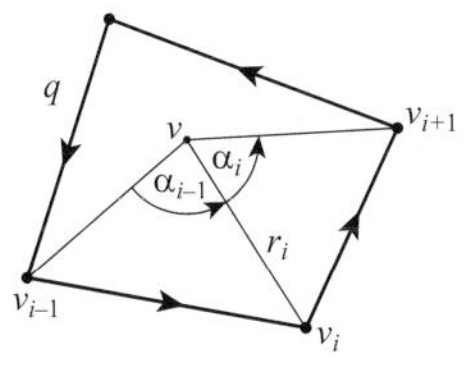

Figure 3: *Notation used for mean value coordinates.*

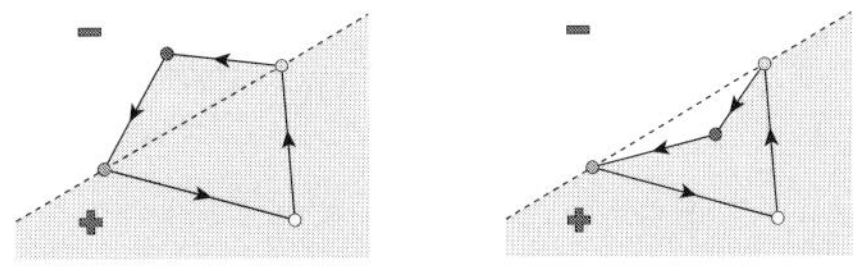

Figure 4: *For a quad with positive orientation, the homogeneous coordinate with respect to the blue vertex is positive in the shaded region and likewise for the other vertices.*

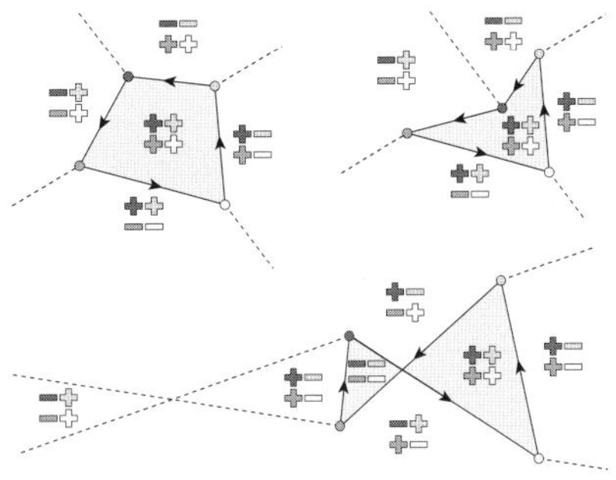

Figure 5: *The homogeneous coordinates all have the same sign inside the quad and the sign depends on the local orientation. Outside the quad, the signs are mixed.*

choices of functions μ_i there exist points v in the interior of such quads for which the denominator in (7) is zero, so that the $\lambda_i(v)$ are not well-defined. This happens, for example, in the case of *Wachspress coordinates* [Wac75, MLBD02].

One choice that does not suffer from this drawback, and has the additional advantage of a modest computational cost (Section 4.2), is the set of *mean value coordinates* [Flo03]. They are defined by

$$\mu_i(v) = \frac{\tan(\alpha_{i-1}(v)/2) + \tan(\alpha_i(v)/2)}{r_i(v)}, \quad (8)$$

where $\alpha_i(v)$ is the *signed* angle in the triangle $[v, v_i, v_{i+1}]$ at v and $r_i(v) = \|v - v_i\|$ is the distance between v and the i-th vertex (see Figure 3).

It follows directly from the definition (8) that if the screen quad q is positively oriented (i.e. counterclockwise), then μ_i is positive inside q and on the left side of the (oriented) line through v_{i-1} and v_{i+1} outside the quad (see Figure 4). If q is negatively oriented, then all signs are reversed, but in any case the key property is that the μ_i all have the same sign inside q and mixed signs outside, even in the case of self-intersecting quads (see Figure 5).

Therefore, the normalization in (7) is always well-defined for a point v inside the quad and its barycentric coordinates $\lambda_i(v)$ are all positive while they have mixed signs if v lies outside. Moreover, the λ_i have the Lagrange property and are linear along the edges, except at the intersection vertex of a self-intersecting quad.

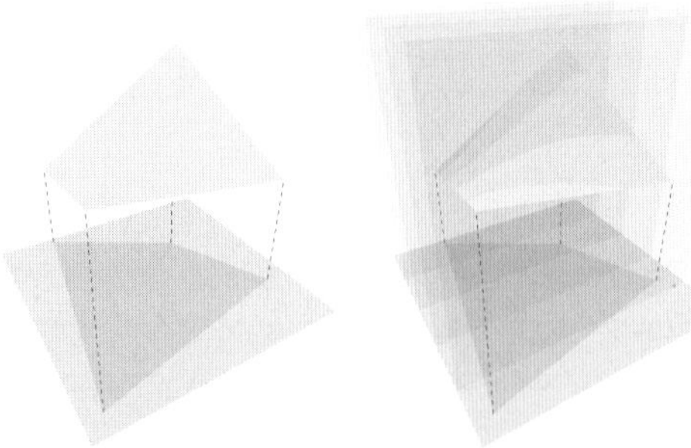

Figure 6: *A flat shaded quad (left) and its intersection with a sequence of parallel semi-transparent planes (right).*

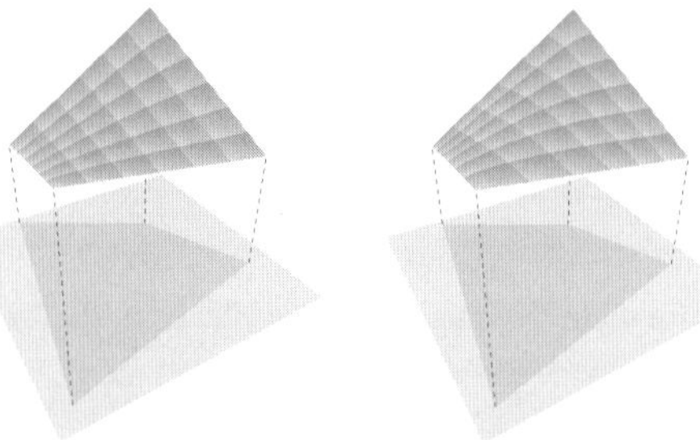

Figure 7: *Texture mapping with standard (left) and perspectively correct interpolation (right).*

3.1. Perspective correction

The simplest rendering method for a triangle is *flat shading* and using the same idea for a quad gives the expected result: a four-sided region that is filled with a constant colour. But the situation gets more exciting if we interpolate the depth values z_i of the four vertices with Equation (4) and assign the resulting depth $z(v)$ to each of the generated fragments. By intersecting the quad with parallel planes we see that this turns the quad into a smooth surface whose intersections with planes are generally curved (see Figure 6).

A nice observation is that we can explicitly describe this surface. If we take any point v inside the quad, then what we really see on screen at this position is the 3D point

$$S(v) = \lambda_0'(v)V_0 + \lambda_1'(v)V_1 + \lambda_2'(v)V_2 + \lambda_3'(v)V_3, \qquad (9)$$

where the *perspective coordinates* $\lambda_i'(v)$ are defined as

$$\lambda_i'(v) = \frac{\lambda_i(v)}{w_i} \Big/ \sum_{j=0}^{3} \frac{\lambda_j(v)}{w_j} = \frac{\mu_i(v)}{w_i} \Big/ \sum_{j=0}^{3} \frac{\mu_j(v)}{w_j}. \qquad (10)$$

with w_i from (1). In fact, if we write $S(v)$ in its homogeneous form and project it to screen space by first multiplying the transformation matrix $\mathbf{M}$ as in (1) and then dividing by the last component, then it is easy to show that this point maps exactly back to v with depth $z(v)$.

Apart from its theoretical value, this observation also has a practical one as it can be used for a *perspectively correct* interpolation of attributes inside the screen quad. Consider, for example, the edge midpoint $v = (v_0 + v_1)/2$ in screen coordinates. For this point we have $\lambda_0(v) = \lambda_1(v) = 1/2$ and $\lambda_2(v) = \lambda_3(v) = 0$ and it follows from (9) that the corresponding 3D surface point is $S(v) = (1-\sigma)V_0 + \sigma V_1$ with $\sigma = w_0/(w_0 + w_1)$ usually not equal to $1/2$. Since attribute interpolation should be linear along the edges of the 3D quad Q, the correct value at v is $(1-\sigma)a_0 + \sigma a_1$, in contrast to the attribute $(a_0 + a_1)/2$ that we get with the interpolation in (4). However, we can still use this equation for perspectively correct interpolation if we interpolate the attributes a_i/w_i and divide the result by the interpolation of the values $1/w_i$,

$$a(v) = \sum_{i=0}^{3} \lambda_i(v) \frac{a_i}{w_i} \Big/ \sum_{i=0}^{3} \lambda_i(v) \frac{1}{w_i}. \qquad (11)$$

Note that this is just a restatement of Equation (9) with the V_i replaced by a_i and that this "trick" is well-known for triangles. An example that illustrates the effect of perspectively correct interpolation for a quad is shown in Figure 7.

3.2. Points behind the eye

So far, we have tacitly assumed that all vertices lie *in front* of the point of view (or *eye*), in other words, that all w_i from Equation (1) are positive. We will now show how to adapt our technique to work even in the case in which some vertices lie *behind* the eye.

For a better understanding, consider what happens if we move one vertex of a quad towards the eye and beyond while the other vertices remain fixed (see Figure 8). As V_1 approaches the eye-plane, the projected vertex v_1 leaves the screen at the bottom because w_1 converges to zero and therefore y_1 diverges to $-\infty$. The moment we push V_1 beyond this plane, w_1 becomes negative, y_1 jumps from $-\infty$ to $+\infty$, and if we keep moving, v_1 will eventually re-appear at the top of the screen.

In analogy to triangles, we have to render the *external* screen quad (see Figure 9) in this situation and in general whenever the vertices V_i lie on different sides of the eye-plane so that the w_i have different signs. This implies that we must consider the external angles and distances for the computation of the homogeneous coordinates in Equation (8).

For example, in the situation at the bottom of Figure 8, α_0' and r_1' should be used instead of α_0 and r_1 (cf. Figure 3). While it is clear that $\alpha_0' = \pi + \alpha_0$, it is less obvious how to compute the external distance, but it can be shown that $r_1' = -r_1$ gives the correct result, i.e. the barycentric coordinates that we compute with these values still have all the desired properties (Lagrange property at the vertices, linearity along the edges, and positivity inside the quad).

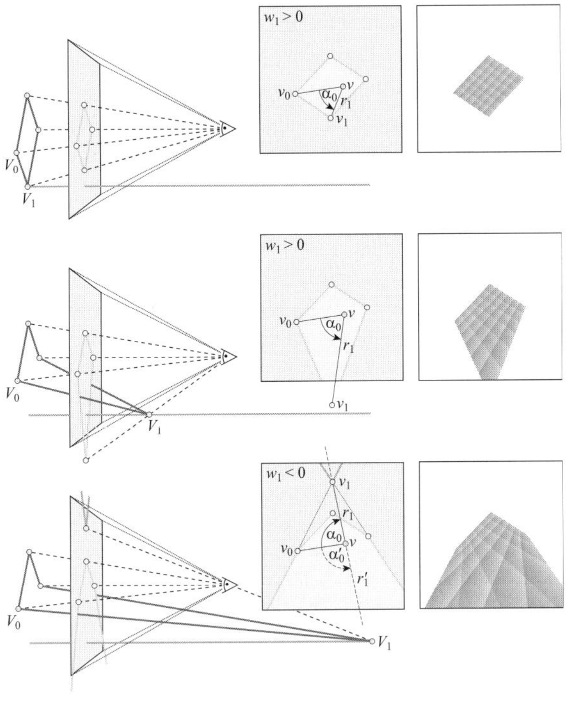

Figure 8: *If a vertex of the 3D quad moves behind the eye (left), then the screen quad becomes an external quad (middle), but the barycentric interpolation still works (right) if we use the appropriate external angles and distances.*

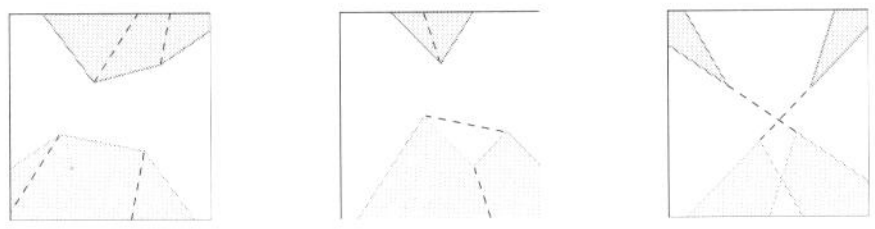

Figure 9: *Examples of the exterior shapes that the configurations in Figure 2 can assume under perspective projection. Red parts lie behind the eye and will be clipped.*

4. Implementation

Traditional hardware implementations for the rasterization of triangles are based on a simplified version of the general algorithm for scan-converting arbitrary polygons. This algorithm intersects successive scan-lines with the polygon and fills the resulting *spans* of adjacent interior pixels from left to right [FvDFH90]. This procedure simplifies a lot in the case of triangles because the number of spans is always one and it further allows to compute the attribute interpolation (4) or (11) with incremental updates on-the-fly. The general algorithm can also be simplified considerably for quadrilaterals as the maximal number of spans is two and the book-keeping of what is called *active edges* is kept at a minimum. We further explain in Section 4.2 how incremental updates can be used to optimize the computation of the attribute interpolation with mean value coordinates.

4.1. Membership test

The basic scan-line algorithm visits pixels sequentially and a common variant to speed it up is to process several pixels in parallel per clock cycle at the cost that not all of them necessarily lie inside the primitive [Pin88]. The *PixelPlanes* architecture [FIP*89] takes this idea to the extreme and completely abandons the scan-line idea by processing all pixels of a 128×128 patch simultaneously using a SIMD array. Another alternative to the scan-line algorithm is to visit pixels along a space-filling Hilbert curve in a hierarchical fashion [MWM01].

All these approaches have in common that they require an efficient test to determine whether a pixel lies inside or outside the primitive. As noticed in Section 3, this test can be implemented by checking the signs of the barycentric coordinates $\lambda_i(v)$, but especially for quads a more efficient solution is desirable.

For triangles the membership is usually tested with the help of *edge functions* [FIP*89, Pin88]. An edge function $E_{ij}(v)$ is defined to be positive (true) if and only if v lies on the left side of the line through v_i and v_j and the simplest function that fulfils this requirement is the signed area of the triangle $[v, v_i, v_j]$. It follows that a pixel at position v is inside the triangle $[v_0, v_1, v_2]$ if and only if the three edge functions E_{01}, E_{12}, and E_{20} have a common sign and the sign itself further indicates whether the pixel is front- or back-facing.

For quads there is a similar strategy that requires to evaluate the four edge functions E_{01}, E_{12}, E_{23}, and E_{30} that correspond to the quad's edges plus a fifth edge function E_{20} that tests against one of the diagonals. A detailed analysis of all the different cases that can occur (see Figure 2) reveals that a pixel is inside the quad and front-facing if and only if either of the boolean expressions

$$\begin{aligned} & E_{20} \wedge E_{01} \wedge E_{12} \wedge (E_{23} \vee E_{30}), \\ & (\neg E_{20}) \wedge E_{23} \wedge E_{30} \wedge (E_{01} \vee E_{12}) \end{aligned} \tag{12}$$

is true and that it is inside and back-facing if the same holds after negating all edge functions. An elegant way to handle exterior quads (see Figure 9) is to evaluate the homogeneous variants of the edge functions as suggested by Olano and Greer [OG97] in the case of triangles.

4.2. Barycentric coordinates

For every pixel that has passed the membership test we now want to interpolate attributes either in the standard way (4) for the depth value or with perspective correction (11) for any other attribute. Both interpolations require to compute and normalize the mean value coordinates (8) which can be simplified using the identity

$$\tan\left(\frac{\alpha_i}{2}\right) = \frac{r_i r_{i+1} - D_i}{A_i}$$

where r_i is the length of the vector $s_i = v_i - v$, the value D_i denotes the dot-product between s_i and s_{i+1}, and A_i is twice the signed area of the triangle $[v, v_i, v_{i+1}]$. A similar identity holds for exterior angles $\alpha_i' = \pi + \alpha_i$,

$$\tan\left(\frac{\alpha_i'}{2}\right) = -\cot\left(\frac{\alpha_i}{2}\right) = \frac{-r_i r_{i+1} - D_i}{A_i},$$

and substituting these formulas as well as any occurrence of exterior distances $r_i' = -r_i$ into Equation (8) leads to the following pseudo-code that handles all cases correctly:

barycentric coordinates (point2D v)

$$\begin{array}{ll}
s_i = v_i - v & i = 0,\dots,3 \\
A_i = s_i \times s_{i+1} & i = 0,\dots,3 \\
D_i = s_i \cdot s_{i+1} & i = 0,\dots,3 \\
r_i = \|s_i\| \cdot \operatorname{sign}(w_i) & i = 0,\dots,3 \\
t_i = (r_i r_{i+1} - D_i)/A_i & i = 0,\dots,3 \\
\mu_i = (t_{i-1} + t_i)/r_i & i = 0,\dots,3 \\
\Sigma = \mu_0 + \mu_1 + \mu_2 + \mu_3 & \\
\lambda_i = \mu_i/\Sigma & i = 0,\dots,3
\end{array}$$

return $(\lambda_0, \lambda_1, \lambda_2, \lambda_3)$

This code can be further simplified in a hardware implementation because A_i is linear and the values D_i and r_i^2 vary quadratically in v. Thus, all of them can be computed efficiently with incremental updates from pixel to pixel and the vectors s_i are no longer needed then. Analogously to triangles, the initial values are generated in a *quad setup* phase that precedes the actual rasterization.

4.3. Back-face culling and clipping

For a triangle, back-face culling can be decided for the entire primitive and the same holds for quads except if the screen quad self-intersects (see Figure 2). In that situation, the shape decomposes into two triangles, one of which is front-facing while the other is back-facing and should be culled. Note that span coherence allows discarding entire spans at a time in a scan-line implementation. However, the best implementation of back-face culling depends heavily on the chosen rasterization technique.

Similar considerations hold for clipping against the z_{near}- and z_{far}-planes. An ad hoc approach is to discard fragments with a depth value outside the interval $[z_{\text{near}}, z_{\text{far}}]$, but depending on the specific architecture, there may be ways to clip them earlier in the pipeline. Note that clipping against the sides of the viewing frustum usually comes for free by producing only fragments that lie inside the viewport.

5. Results

Although our method is intended to be implemented as a hardware rasterizer, in order to test it and have a visual insight to the results obtained with it, we resorted to an hybrid software/hardware implementation that runs on current triangle rasterizing GPUs. All rendering examples shown in this paper have been produced with this implementation.

5.1. Hybrid implementation

In our test implementation, all vertex processing (including the projection) has been lifted from the vertex shader to the CPU, while the computation of the barycentric coordinates moved down from the hypothetical rasterizer to the fragment shader. The result is inefficient compared to a real hardware rasterizer, but it is good enough to render simple quad meshes in real-time.

To be specific, we first project all vertices according to Equation (1), record the values v_i, z_i, and w_i, and store them into the current rendering state so that we can access them in the fragment shader later. The original vertex attributes a_i (e.g. per-vertex normals) are similarly stored in this state.

We then render the quad by simply splitting it into two triangles. Each fragment that is produced by the triangle rasterizer is processed by a fragment program that computes the barycentric coordinates λ_i from the fragment's screen position v and from the points v_i as shown in the pseudo-code above and carries out the necessary attribute interpolations as mentioned in Section 4.2. Interpolated attributes are then processed as usual to compute the final colour value.

By rendering the two triangles we are sure to generate all fragments inside the quad plus some additional ones in some cases (for concave and self-intersecting shapes) that lie outside. The latter ones are detected and killed in the fragment program by checking the signs of the barycentric coordinates as mentioned in Section 3.

5.2. Properties and discussion

The improvement of our method over the currently used diagonal split is remarkable and the examples in Figures 1, 10, and 11 emphasize that the quality matches more or less the one obtained by an expensive refinement of the quad with bilinear interpolation in 3D. The quality mainly stems from the smoothness of the mean value coordinates that guarantees a C^∞-continuous interpolation inside the quad.

Another important characteristic of our rendering technique is that it is highly compatible with triangle rendering. The linearity of the attribute interpolation along edges ensures C^0-continuity across adjacent quads or between neighbouring quads and triangles; crease angles can still be obtained as usual by assigning different normals at the vertices shared by two adjacent quads (or triangles); and depth values are computed consistently with other primitives. Furthermore, the only part of the graphics pipeline that is affected is the rasterizer and any per-vertex and per-fragment processing remains unchanged.

Another advantage is that our quads are compliant with the *OpenGL* specification [SA03] and only the API *implementation* of the primitives created in GL_QUAD or GL_QUAD_STRIP mode would have to be replaced. Direct3D syntax, however, would require an extension to include the "quad" as a new "primitive type".

Figure 10: *Visual comparison of different methods to render a single textured quad. We added a shadow and dotted vertical lines to suggest the actual 3D geometry of the quad. From left to right: the two diagonal spits, a 20×20 refinement of the quad using bilinear interpolation, and our proposed method that renders the quad as an atomic primitive. Each column shows the same quad seen from different points of view, resulting in the three possible projected shapes (cf. Figure 2).*

The cost of the proposed technique is an increased complexity of the rasterizer. The part of the ARB fragment program that computes the barycentric coordinates in our hybrid test implementation (see Section 5.1) is only 19 instructions long and this suggests that a hard-wiring in a rasterizer would not be too demanding.

A potential drawback of our method is the view dependency of the result as the underlying 3D shape of the rendered surface (see Section 3.1) varies according to the current viewing direction (except for planar quads). This is unavoidable because we always fill the interior of a 2D screen quad and a non-planar 3D surface that projects to a quadrilateral shape from all viewing directions does not exist. On the other hand, this guarantees that every pixel corresponds to exactly one surface point and therefore enables a straightforward rendering of two-sided quads. The effect of view dependency can be undesirable at times and becomes especially visible for textured non-planar quads that cover big portions of the screen. But it is almost invisible for quad meshes with higher resolution for which the diagonal split still creates ugly discontinuities (see Figure 11).

Acknowledgements

This work was supported in part by the Deutsche Forschungsgesellschaft (DFG HO-2457/1-1) and the projects *ViHAP3D* (EU IST-2001-32641) and *MACROGeo* (FIRB-MIUR RBAU01MZJ5). Thanks to Michael Floater, Paolo Cignoni, and our reviewers for their helpful suggestions.

References

[Duf79] DUFF T.: Smoothly shaded renderings of polyhedral objects on raster displays. *ACM SIGGRAPH Computer Graphics 13*, 3 (1979), 270–275. Proceedings of SIGGRAPH '79.

[FHK04] FLOATER M. S., HORMANN K., KÓS G.: A general construction of barycentric coordinates over convex polygons. *Advances in Computational Mathematics* (2004).

[FIP*89] FUCHS H., ISRAEL L., POULTON J., EYLES J., GREER T., GOLDFEATHER J., ELLSWORTH D., MOLNAR S., TURK G.,

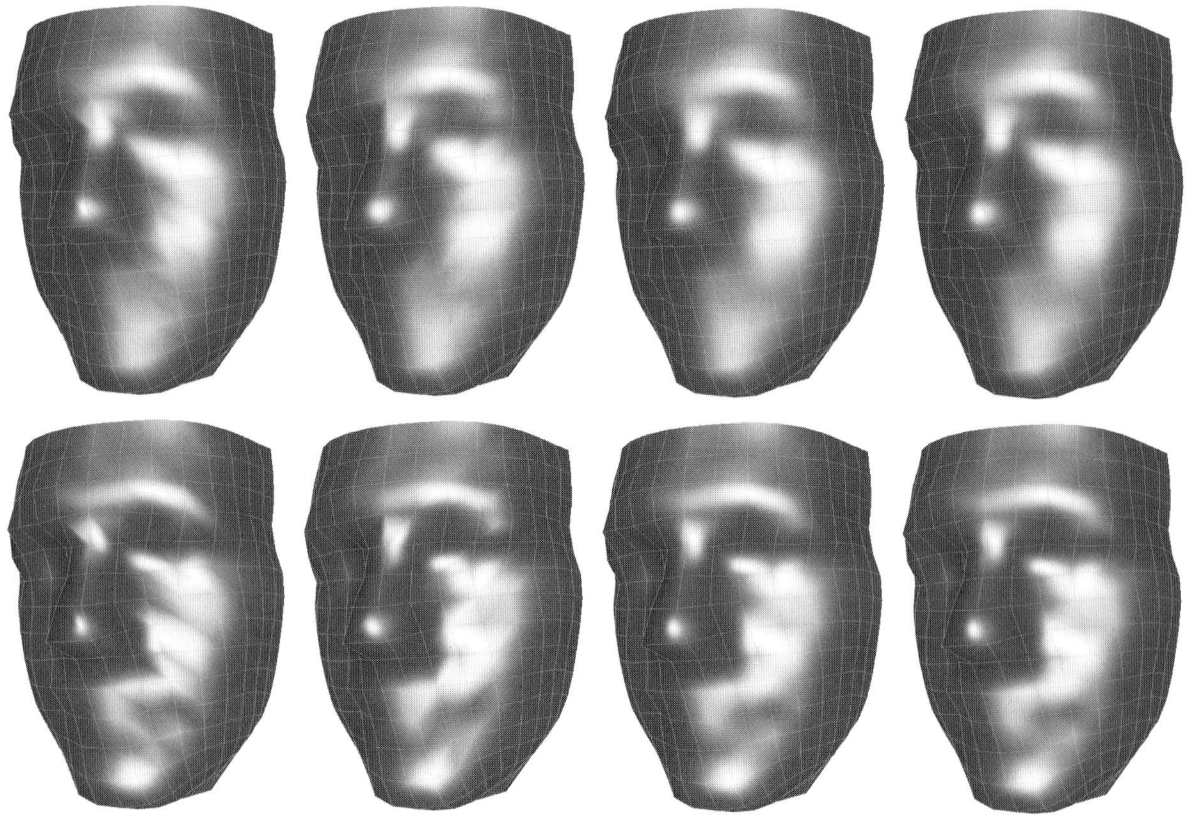

Figure 11: *Visual comparison of different methods to render a simple Phong-shaded quad mesh. The over-imposed wireframe shows the connectivity. From left to right: the two diagonal spits, a 20×20 refinement of the quads using bilinear interpolation, and our proposed method that renders the quads as atomic primitives. The per-vertex normals that were in the examples on top were computed from the 3D geometry. In the bottom row, we used the normals that were originally defined for the model which yield a more accurate description of the actual shape, but also make the artefacts of the diagonal split more evident.*

TEBBS B.: Pixel-planes 5: a heterogeneous multiprocessor graphics system using processor-enhanced memories. *ACM SIGGRAPH Computer Graphics 23*, 3 (1989), 79–88. Proceedings of SIGGRAPH '89.

[Flo03] FLOATER M. S.: Mean value coordinates. *Computer Aided Geometric Design 20*, 1 (2003), 19–27.

[FvDFH90] FOLEY J. D., VAN DAM A., FEINER S., HUGHES J.: *Computer Graphics: Principles and Practice*. Addison-Wesley, 1990.

[MLBD02] MEYER M., LEE H., BARR A. H., DESBRUN M.: Generalized barycentric coordinates on irregular polygons. *Journal of Graphics Tools 7*, 1 (2002), 13–22.

[MWM01] MCCOOL M. D., WALES C., MOULE K.: Incremental and hierarchical Hilbert order edge equation polygon rasterization. In *Proceedings of the 2001 Workshop on Graphics Hardware* (2001), pp. 65–72.

[OG97] OLANO M., GREER T.: Triangle scan conversion using 2D homogeneous coordinates. In *Proceedings of the 1997 Workshop on Graphics Hardware* (1997), pp. 89–95.

[Pin88] PINEDA J.: A parallel algorithm for polygon rasterization. *ACM SIGGRAPH Computer Graphics 22*, 4 (1988), 17–20. Proceedings of SIGGRAPH '88.

[SA03] SEGAL M., AKELEY K.: The OpenGL graphics system: A specification (version 1.5), Oct. 2003. http://www.opengl.org/documentation/specs/version1.5/glspec15.pdf.

[Wac75] WACHSPRESS E. L.: *A Rational Finite Element Basis*. Academic Press, 1975.

Graphics Hardware (2004)
T. Akenine-Möller, M. McCool (Editors)

A Hierarchical Shadow Volume Algorithm

Timo Aila*† Tomas Akenine-Möller‡

*Helsinki University of Technology/TML, †Hybrid Graphics, Ltd., ‡Lund University

Figure 1: *From left to right: Correct shadows, low-resolution shadows (sample per* 8×8 *pixel tile), tiles that may contain a shadow boundary (green/light gray), and an image showing accurately processed boundary tiles (darker gray) and data copied from the low-resolution shadows (lighter gray). If a tile contains a shadow boundary (3rd image), the corresponding low-resolution shadow data is more or less random. This is corrected by applying per-pixel rasterization to boundary tiles.*

Abstract

The shadow volume algorithm is a popular technique for real-time shadow generation using graphics hardware. Its major disadvantage is that it is inherently fillrate-limited, as the performance is inversely proportional to the area of the projected shadow volumes. We present a new algorithm that reduces the shadow volume rasterization work significantly. With our algorithm, the amount of per-pixel processing becomes proportional to the screen-space length of the visible shadow boundary instead of the projected area. The first stage of the algorithm finds 8×8 *pixel tiles, whose 3D bounding boxes are either completely inside or outside the shadow volume. After that, the second stage performs per-pixel computations only for the potential shadow boundary tiles. We outline a two-pass implementation, and also describe an efficient single-pass hardware architecture, in which the two stages are separated using a delay stream. The only modification required in applications is a new pair of calls for marking the beginning and end of a shadow volume. In our test scenes, the algorithm processes up to* 11.5 *times fewer pixels compared to current state-of-the-art methods, while reducing the external video memory bandwidth by a factor of up to* 17.1.

Categories and Subject Descriptors (according to ACM CCS): I.3.7 [Computer Graphics]: Three-Dimensional Graphics and Realism—Shadowing I.3.1 [Computer Graphics]: Hardware Architecture—Graphics Processors

1. Introduction

Rendering of shadows is a key ingredient in computer-generated images. Shadows both increase the level of realism and provide information about spatial relationships of objects. For real-time rendering, variants of the shadow mapping algorithm [Wil78] and the shadow volume algorithm [Cro77] are among the most popular techniques. Shadow mapping is an image-based technique that often suffers from aliasing problems. The shadow volume algorithm does not have these issues, but has been criticized for its excessive use of fillrate. Shadow volumes are extruded from shadow casters to the attenuation range of a light source. Even relatively small objects, such as characters, often create shadow volumes that cover a significant portion of the screen.

The main motivation for our work is that when a shadow volume is rasterized, only a small portion of the processed pixels typically define the footprint of the shadow, and the resulting shadow often consists of large homogeneous areas. Clearly, there appears to be ample room for improvement in terms of performance.

We present a two-stage algorithm for hierarchical shadow volume rendering. The first stage finds 8×8 pixel tiles that are either completely outside or inside shadow volumes of individual objects, i.e., fully lit or fully in shadow. The results of this stage are valid after the entire shadow volume has been processed. Therefore the shadow volumes need to be either submitted twice, or the shadow volume triangles need to be delayed in the rasterization pipeline by using a delay stream [AMN03]. After the first stage is complete, the second stage needs to perform per-pixel shadow volume rasterization only for the boundary tiles.

Compared to existing acceleration techniques [NVI03], the results show that our algorithm processes 2.8–11.5 times fewer pixels, and reduces the related bandwidth usage by a factor of 2.4–17.1. We firmly believe that these results justify the proposed hardware modifications, because without our new algorithm, up to over 90% of the total rendering time was spent in shadow volume rasterization, as discussed in Section 6.

Although this paper focuses on hard shadows, the technique can be used in different contexts as well. The soft shadow volume algorithm [AAM03] can be accelerated, since it contains a hard shadow volume pass, and its culling optimization [ADMAM03] can benefit from our new method.

The rest of the paper is organized as follows. We begin by reviewing related work in Section 2, and proceed by describing our algorithm in Section 3. Section 4 outlines two implementations, and Section 5 presents the details regarding a hardware implementation, followed by our test results and discussion in Section 6. Finally, we offer conclusions and some thoughts on future work in Section 7.

2. Previous Work

In this section we focus on presenting previous work on real-time generation of hard shadows using graphics hardware, and briefly cover the use of delay streams. For a general survey on shadow generation, consult Woo et al. [WPF90]. A more recent survey covering real-time algorithms is also available [HM01].

Shadow mapping [Wil78] generates a depth buffer from the point of view of the light source. This image, called a *shadow map*, is a discretized representation of the scene geometry as seen by the light source. A pixel is in shadow if its depth value, transformed into the light-space, is greater than the corresponding depth value in the shadow map. The discretization often causes aliasing artifacts such as jagged shadow boundaries and incorrect self-shadowing. Sen et al. [SCH03] improve the accuracy of shadow boundaries, and give references to other work regarding the aliasing issues.

A shadow volume consists of three parts. The *light cap* is the shadow casting object itself, the *dark cap* closes the far end of the shadow volume at infinity or at the attenuation range of the light source. The *side quads*, extruded from the silhouette edges of the shadow caster, connect the two caps into a closed volume.

In a hardware accelerated variant [Hei91] of the original shadow volume algorithm [Cro77], the scene is first rendered using ambient lighting. In the second pass, the shadow volumes are rendered into the stencil buffer so that visible pixels of the front-facing and back-facing triangles increment and decrement the stencil values. This creates a shadow mask in the stencil buffer, where values greater than zero indicate that a pixel is in shadow. In a third pass, the scene is rendered with full lighting, and the per-pixel shadow terms of the stencil buffer are used to mask out shadowed pixels. This method is commonly known as the *Z-pass* shadow volume algorithm. In contrast to shadow mapping, the shadow volume algorithm does not suffer from aliasing problems.

Problems occur when the camera is inside a shadow volume, and a more robust solution is used nowadays. The *Z-fail* algorithm has been independently described by Bilodeau & Songy and by Carmack, as discussed by Everitt and Kilgard [EK02]. The depth test is simply reversed when rendering the shadow volumes, so that only the triangles that are behind the contents of the Z-buffer affect the stencil buffer. Two solutions are presented for the cases where a shadow volume intersects the far plane of the view frustum.

Shadow Volume Optimizations Two-sided stencil test [EK02] halves the geometry processing requirements of shadow volumes. NVIDIA's UltraShadow [NVI03] allows the programmer to define the minimum and maximum depth values for a shadow volume. A pixel cannot be inside the shadow volume if the Z-buffer value of the pixel is outside the defined *depth bounds*. Therefore a significant portion of shadow volume rasterization work can potentially be avoided when the shadow receivers are outside the depth bounds. As an optimization, the bounds can be clamped to bounding volumes of local light sources or to scene geometry, e.g., walls of a room. Depth bounds testing is effective when the shadow volume is approximately perpendicular to the viewing direction. However, with other orientations the bounds may cover a major part of the scene and the efficiency degrades. Also, the testing does not accelerate the rendering of shadowed regions.

Lengyel [Len02] uses the scissor test to limit the rasterization work to portions of the screen bounded by the influence region of the light source. McGuire et al. [MHE*03] present a framework that combines depth bounds and Lengyel's scissoring with new optimizations. They introduce an algorithm for creating the dark cap with a minimum number of triangles, and thus in terms of triangle count the light cap becomes the most complex part of a shadow volume. The Z-pass algorithm does not need the light cap, and thus it can be faster to use Z-fail only for the shadow volumes that may intersect the viewport [EK02]. Lloyd et al. [LWGM04] clamp shadow volumes so that the rasterization is limited to regions that contain shadow receivers. They also use occlusion queries to avoid casting shadows from objects that are entirely in shadow. All optimizations listed above can also be used in conjunction with our algorithm.

Hybrid Algorithms McCool [McC00] uses a shadow map and image-space edge detection to reconstruct the shadow volume of a set of objects. This shadow volume does not contain any overlapping volumes, which means that a lot of redundant rasterization work can be avoided. However, the creation of the shadow volume is rather complex and slow, so the technique is not suitable for dynamic scenes. Also, due to the discrete resolution of a depth map and the use of an edge detection algorithm, robustness issues arise.

Govindaraju et al. [GLY*03] use a combination of shadow maps and object-space shadows on a three-PC system to generate hard shadows in complex environments. Their optimizations for limiting the number of shadow casters and receivers could also be used for accelerated rendering of shadow volumes.

Chan and Durand [CD04] first determine the boundary pixels of shadow regions from a low resolution shadow map. The boundary pixels are then processed accurately using shadow volumes, while the rest of the pixels are handled with the shadow map.

Occlusion Culling Meissner et al. [MBGS01] perform two-pass visibility driven rasterization by first marking the tiles that are fully hidden and then skipping the subsequent rasterization to those tiles. Our algorithm bears some similarities to this technique.

The delay stream [AMN03] is a hardware mechanism that delays triangles in the rasterization pipeline. It can be used to generate Z-buffer information early in the pipeline, and after that the triangles are pushed into the delay stream. At a later stage, the triangles are popped, and at that time it is likely that enough Z-buffer information has been accumulated, so that many additional triangles can be culled.

3. Hierarchical Shadow Volume Algorithm

The following explanation of our new algorithm assumes that the frame buffer has been divided into 8×8 pixel tiles. For each tile, the z_{min} [AMS03] and z_{max} [Mor00] of the Z-buffer are maintained. The vertical and horizontal bounds of a tile, along with its z_{min} and z_{max} define a three-dimensional axis-aligned box in screen space. After rasterization of the scene, the box contains all the visible geometry, and is therefore a 3D bounding box of the tile.

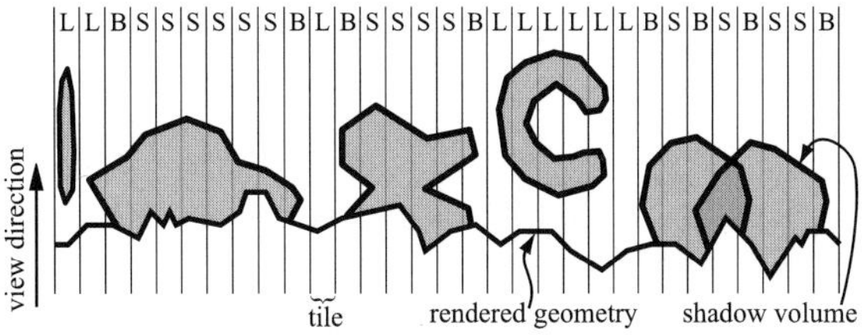

Figure 2: *This 2D illustration of our algorithm assumes the Z-fail algorithm, and thus only the parts of the shadow volumes that are behind the rendered geometry can affect the stencil buffer. The boundary tiles (B) contain intersections of rendered geometry and shadow volume triangles. A visible shadow boundary can only exist inside the B-tiles, and thus only these tiles need to be rasterized accurately. For all other tiles, it is sufficient to execute the shadow volume algorithm for a single point inside the tile, because the result is the same for the whole tile. The L-tiles are fully lit and the S-tiles are fully in shadow.*

3.1. Geometrical Explanation

Our algorithm builds on the fundamental observation that a visible shadow boundary can be inside a tile only if at least one shadow volume triangle intersects with the 3D bounding box of the tile. Two facts guarantee that the approach works. First, shadow volumes are closed by definition. Second, the triangles defining a shadow volume indicate transitions between light and shadow, i.e., if there are no intersections inside a tile, the tile cannot contain a shadow boundary. Hence, it is sufficient to execute the shadow volume algorithm for a single, arbitrarily chosen point inside a non-boundary tile in order to determine the shadow term for all pixels inside the tile. Figure 2 illustrates the idea in two dimensions.

Intuitively the shadow volume algorithm corresponds to shooting a ray through a point inside the tile, and counting the number of intersected frontfacing and backfacing triangles. A key to our method is that depending on which point is chosen, the number of intersections along the ray can change, but the net result is always the same. As an example, consider the leftmost shadow volume in Figure 2 that resides completely inside a single tile. Depending on which point (i.e., vertical ray in Figure 2) we choose inside the tile, the shadow volume may either be completely missed, or the test point will register one back-facing and one front-facing triangle, which cancel each other. The correct result is obtained in both cases: the shadow volume does not contribute to the visible shadow, and can be culled.

3.2. Algorithm Outline

Our algorithm executes in two stages. The process begins when a marker, BSV (Beginning of Shadow Volume), is encountered. First, the screen-space tiles are classified to be either fully lit, fully in shadow, or potentially containing a shadow boundary. This classification depends on the shadow volume as a whole, and remains incomplete until the corresponding `ESV` (End of Shadow Volume) marker is received. When the classification is complete, it can be exploited for efficient rasterization, as we will see in Section 4.

In general there are a number of shadow casters per light source, and in the following explanation each shadow casting object creates a separate shadow volume. Furthermore, the depth buffer has been initialized by rendering the scene geometry. The rest of this section explains the two stages of our algorithm in detail without being limited to any specific implementation.

3.2.1. Stage 1: Low-Resolution Shadows and Tile Classification

The tile classification is performed by using a per shadow volume (PSV) buffer, which stores for each tile an 8-bit `stencil` value and a Boolean `boundary`. Initially `boundary=false` and `stencil=`S_{clear}, where S_{clear} is the stencil buffer clear value. Without loss of generality, the rest of this paper assumes that stencil values greater than S_{clear} indicate shadow.

If a triangle intersects with the 3D bounding box of a tile, there is a potential shadow boundary inside the tile. Such boundary tiles are marked by setting their `boundary` to `true`. The intersection needs to be computed in a conservatively correct manner, i.e., at least all the actual intersections must be detected. Notice that any tile can be classified as a boundary tile without introducing visual artifacts.

If none of the shadow volume triangles intersect with the 3D bounding box of a tile, the entire tile is either fully lit or fully in shadow. Thus the classification can be carried out by executing the shadow volume algorithm for a single, arbitrarily chosen sampling point inside the tile. The shadow volume algorithm updates the `stencil` of the tile by using the defined stencil and depth modes.

After an entire shadow volume has been processed, the corresponding tile classifications are ready. If the `boundary` bit of a tile is set, per-pixel stencil rasterization is needed in order to determine the exact shadow boundary. Otherwise the rasterization can be skipped, because the entire tile is either in shadow (`stencil>`S_{clear}) or lit with respect to the current shadow volume.

The classification allows alternating between Z-pass and Z-fail algorithms, and does not pose limitations on the shadow volume topology or on the ordering of the triangles.

3.2.2. Stage 2: Stencil Buffer Updates

When the `ESV` marker has emerged from Stage 1, the shadow volume triangles are rasterized according to the tile classifications in the corresponding PSV-buffer. The following three cases are possible:

1. Tile is fully inside the shadow volume (S-tiles in Figure 2). All pixels in the tile are set to be in shadow.
2. Tile is fully outside the shadow volume (L-tiles). No updates needed.
3. The shadow volume intersects the 3D bounding box of the tile (B-tiles). Per-pixel stencil operations are performed, as requested by the application.

4. Implementing the Algorithm

Our algorithm can be implemented in various ways. We first introduce a hierarchical variant of the stencil buffer in order to make the stencil buffer updates in Stage 2 more efficient. We then briefly outline a two-pass implementation, and finally explain a more efficient single-pass implementation using a delay stream. Both implementations use the hierarchical stencil buffer.

4.1. Hierarchical Stencil Buffer

We propose a two-level version of the stencil buffer. For each 8×8 pixel tile, the minimum and maximum stencil buffer values, S_{min} and S_{max}, are maintained in a low-resolution stencil buffer. Furthermore, we propose that the stencil test is carried out as follows. If the result of the stencil test can be determined from $[S_{min}, S_{max}]$, the per-pixel stencil buffer must not be accessed. For example, if $S_{min} = S_{max} = 1$, the per-pixel values are not needed for any of the stencil comparison modes. This opens the possibility of not updating or accessing the per-pixel stencil values unless absolutely necessary.

4.1.1. Optimized Stage 2: Hierarchical Updates

The stencil buffer updates in Section 3.2.2 can be optimized by using the hierarchical stencil buffer:

1. Tile is fully inside the shadow volume. The corresponding entry in the low-resolution stencil buffer is updated. Per-pixel operations to the tile are skipped.
2. Tile is fully outside the shadow volume. No updates needed.
3. The shadow volume intersects the 3D bounding box of the tile. Per-pixel stencil operations are performed, as requested by the application. S_{min} and S_{max} are updated according to the per-pixel stencil buffer rasterization.

Per-pixel rasterization and the related video memory bandwidth are limited to the boundary tiles. In general, this results in significant reduction of fillrate requirements, as illustrated in Figure 1. It is worth pointing out that the hierarchical stencil buffer alone would provide only small performance improvements; the two-stage shadow volume algorithm is needed for efficiently exploiting the hierarchical updates.

4.1.2. Final Lighting Pass

After all shadow casters have been processed, the full lighting contribution of the light source is accumulated into the frame buffer so that the shadow terms are read from the hierarchical stencil buffer. Per-pixel stencil values are fetched only for the boundary tiles of the combined shadow area of all shadow volumes. This provides an additional reduction to the memory bandwidth requirements.

4.2. Two-Pass Implementation

Perhaps the simplest implementation of our algorithm requires the application or the device driver to submit each shadow volume twice. The two passes then correspond to Stage 1 and optimized Stage 2. However, this alternative would need a second geometry pass and further modifications to the applications in order to facilitate efficient pipelined execution.

4.3. Single-Pass Implementation Using a Delay Stream

This single-pass implementation uses the optimized Stage 2, whereas Stage 1 remains identical to Section 3.2.1. Between the two stages, the triangles defining a shadow volume are temporarily stored into a delay stream, which is a ring buffer in external video memory [AMN03]. Unlike delayed occlusion culling, our algorithm uses the delay stream only for shadow volumes. Note that the same stream could be used by both algorithms. The `ESV` marker is passed directly from Stage 1 to Stage 2 without going through the delay stream. The rest of this section explains our architecture in detail, and a general overview is given in Figure 3.

4.3.1. Delay Stream

The delay stream should be large enough to hold all the triangles of a single shadow volume in order to delay the stencil buffer rasterization up to the point when the classification is finished. Typically the geometry defining a shadow volume consumes only a small amount of memory for two reasons. First, the shadow triangles usually contain only positional information, i.e., no colors or texture coordinates per vertex, and this leads to good compression of the delay stream. Second, for the majority of shadow volumes, only the side quads are needed [MHE*03], and for a mesh of n triangles, the number of silhouette edges can be as low as $O(\sqrt{n})$ [SHSG01].

In certain pathological cases, e.g., when all objects of a scene are presented as a single triangle soup, the allocated delay stream may not be able to store even one entire shadow volume. If this happens, the per-pixel stencil buffer rasterization has to start before the tile classification is complete, and the performance degrades gracefully towards the traditional shadow volume algorithm. Visual artifacts are avoided by treating all tiles as boundary tiles until the classification

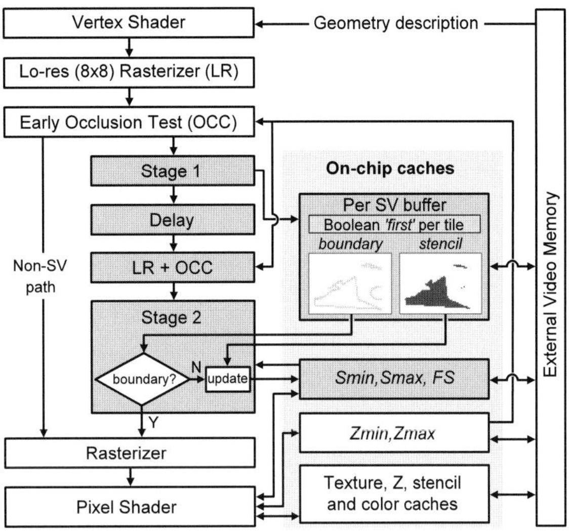

Figure 3: *The positioning and connections of our algorithm inside a modern programmable graphics pipeline. The new units and on-chip caches are marked with gray.*

finishes. After that, the remaining per-pixel rasterization is skipped for tiles that were classified to be fully in shadow. The upper bound of per-pixel stencil rasterization work is equal to the traditional shadow volume algorithm.

4.3.2. Further Optimizations

Usually there are multiple objects casting shadows from a light source. When the contribution of a shadow volume is added to the stencil buffer, the overall area covered by the shadow grows monotonically. For example, a tile that has been previously classified as being fully in shadow cannot be downgraded to a boundary tile. This is especially useful for scenes that have a high depth complexity, such as a city scene.

We perform this check by using a full shadow (FS) buffer, which stores one bit per tile. This optimization needs a separate FS-buffer because the contents of the stencil buffer are modified during the rasterization of a shadow volume, and are thus in a meaningful state only between two shadow volumes. Therefore we must update the FS-buffer exactly once per shadow volume. For this purpose, we add a Boolean `first` to each tile in the PSV-buffer, and initialize them to `true` immediately after receiving the BSV in Stage 1. If the `first` bit of a tile is set, the FS bit is set to `true` if $S_{min} > S_{clear}$ and to `false` otherwise. The `first` is then cleared in order to avoid examining the stencil buffer during the processing of a shadow volume. Subsequent rasterization to a tile is skipped if the FS bit is set to `true`.

Appendix A contains pseudo code for Stages 1 and 2. The use of FS-buffer makes the performance of our algorithm slightly dependent on the processing order of shadow volumes. However, we have not optimized our test applications to take this into account.

5. Hardware Implementation of Single-Pass Algorithm

The pipelined processing of shadow volumes deserves special attention. The tile classifications are made individually for each shadow volume, and thus Stages 1 and 2 are generally processing different shadow volumes. Therefore multiple per shadow volume (PSV) buffers are needed. We handle this by allocating in the driver a small number of buffers, according to the screen resolution. The buffers are stored in external video memory and accessed through an on-chip cache.

A PSV-buffer is locked for a particular shadow volume in Stage 1 when `BSV` is encountered. If no buffers are available, the unit stalls. The buffer is released in Stage 2 after `ESV` is received. Only a part of each buffer is generally accessed by a shadow volume, and thus a clear bit [Mor00] per 32×32 pixels provides a fast way of clearing the necessary parts of the PSV-buffers in Stage 1.

5.1. New Hardware Resources

The new hardware resources consist of computational logic in Stages 1 and 2, a delay stream, additional instances of the low-resolution rasterizer and early occlusion test units, and on-chip storage. The amount of new logic is fairly low as can be seen from the pseudo code in Appendix A. The only non-trivial operations are finding the intersection of a shadow volume triangle with the 3D bounding box of a tile, and determining if the triangle is behind the bounding box. However, both of these questions need to be answered already in the preceding "Early Occlusion Test" unit (Figure 3) in order to perform z_{min} and z_{max} culling. Such a unit already exists in most modern graphics chips, and thus the implementation details are omitted from this paper.

The delay stream is a ring buffer in external memory, and the new hardware units are limited to a simple history-based compression [AMN03], which eliminates duplicate vertices and stores only the active attributes of vertices, i.e., the position. The render state changes are also stored, although there are usually very few of them during the rasterization of shadow volumes.

On-chip storage consists of the PSV-buffer cache, FS-buffer and $S_{min,max}$ buffers. Since the tile classification (Section 3.2.1) allows arbitrary positioning of the sample point within the tile, one can let four tiles, placed in a 2×2 configuration, share the same sample point, and thus also the same `stencil` value in the PSV-buffer. In this approach, the sample point is located at the shared corner of the four tiles. The storage for the PSV-buffer is then $2 + 8/4 = 4$ bits per tile. A 2kB cache was used in our tests. The FS-buffer uses one bit per tile, which corresponds to 2.5kB in 1280×1024 resolution.

$S_{min,max}$ can be useful for generic computations using the stencil buffer. However, if they are only used for the purposes of shadow volume rendering, the stencil value can mean only one of two things, namely the pixel being lit or in shadow. With such encoding "1" indicates shadow and "0" means lit, and partial shadow is indicated by $S_{min} = 0$ and $S_{max} = 1$. While this 1-bit representation is adequate for shadows, it imposes two limitations. First, the function deciding what stencil values correspond to shadow should not be changed between constructing and using the stencil buffer. This does not appear to be a serious limitation in practice. Second, the $S_{min,max}$ buffers contain meaningful data for shadow computations, but should be disabled for other computations that perform stencil tests. One way of implementing this is to prefix the stencil test with a new shadow test, which uses $S_{min,max}$. The traditional stencil test is executed only if the shadow test cannot determine the results, e.g., a tile is partially in shadow or the shadow test is disabled. As a result, the size of $S_{min,max}$ can be made as small as 5kB in 1280×1024 resolution. For comparison, the existing $z_{min,max}$ buffers consume 16 times more memory, assuming 16-bit depth values.

6. Results and Discussion

The performance of our single-pass algorithm was benchmarked using walkthroughs in four scenes (Table 1). The complexity of the scenes is rather accurately illustrated by the average number of rendered shadow volume triangles per frame. Only eight were rendered in the Simple scene, and at the other end of the scale, 572k in the Powerplant. The two intermediate scenes, Knights (18k) and Amphitheater (6.6k), include various animated characters. All measurements were done using a single light source, and for Simple scene, Knights, and Powerplant the light source was moving.

Three different variants of the shadow volume algorithm were implemented inside MESA (`www.mesa3d.org`), which is a software renderer with an OpenGL-like interface. The first variant is a brute force approach, the second is based on depth bounds [NVI03], and the third is our hierarchical algorithm. All three utilize z_{min}, z_{max}, and lossless compression for the Z-buffer [Mor00]. The compression is assumed to reduce Z-buffer bandwidth usage by 50%. The depth bounds were computed separately for each shadow casting object. The bounds were initialized with the bounding box of the shadow caster, and then extended to include the intersection of the view frustum and the shadow volume of the bounding box. In Amphitheater shadow volumes are extruded from the local light source to infinity instead of the attenuation range, and thus the tightness of depth bounds could have been further improved by clamping the bounds to the bounding volume of the light source.

The images were rendered at 1280×1024 resolution, and all accesses to $z_{min,max}$, $S_{min,max}$ and FS-buffer were on-chip.

	Simple scene			Knights			Powerplant (Section 16)			Amphitheater		
Algorithm	*BF*	*DB*	*H*	*BF*	*DB*	*H*	*BF*	*DB*	*H*	*BF*	*DB*	*H*
Stencil BW in shadow pass	0.54	0.48	0.03	48.96	18.65	5.90	4329	1798	123	9.59	9.05	1.66
Z-buffer BW in shadow pass	0.05	0.05	0.05	3.00	3.00	2.93	160.3	160.3	115.1	0.97	0.97	0.84
Stencil BW in lighting pass	1.24	1.24	0.02	0.82	0.82	0.11	1.16	1.16	0.01	0.81	0.81	0.07
PSV BW from 2kB cache	-	-	0.005	-	-	0.035	-	-	29.2	-	-	0.19
Delay stream BW	-	-	0.001	-	-	0.43	-	-	15.1	-	-	0.16
Total bandwidth	1.82	1.77	0.10	52.78	22.46	9.40	4490	1959	283	11.37	10.82	2.91
#Pixels in shadow pass (M)	0.28	0.25	0.02	25.67	9.78	3.48	2269	943	99	5.03	4.74	0.94
Ratio in bandwidth		1.03:1	**17.6:1**		2.35:1	**5.61:1**		2.3:1	**16.0:1**		1.05:1	**3.91:1**
Ratio in #processed pixels		1.1:1	**12.7:1**		2.63:1	**7.38:1**		2.4:1	**22.9:1**		1.06:1	**5.35:1**

Table 1: *Average per-frame statistics for brute force* (BF)*, depth bounds* (DB*) and our hierarchical algorithm* (H) *in* 1280 × 1024 *resolution with 32-bit depth buffer and 8-bit stencil buffer. All bandwidth (BW) measurements are shown in megabytes (MB). Ratios are relative to the brute force algorithm.*

The measured total bandwidth in Table 1 refers to the cost of constructing the shadow mask into the stencil buffer, and then fetching the shadow terms in the subsequent illumination pass. Z-fail was used in the Simple Scene, Knights and Powerplant due to its robustness. Adaptive switching between Z-pass and Z-fail could have reduced the number of shadow volume triangles, but we did not implement it because the fillrate measurements are independent of the triangle count, and there is no proof that either Z-pass or Z-fail has better bandwidth usage. However, the Amphitheater is publicly available as a part of an application that readily implements the switching. The size of the delay stream was set to 1MB, and Stage 2 started to process a shadow volume as soon as Stage 1 finished classifying it.

Our algorithm reduced the total bandwidth by a factor of 3.9–17.6 compared to the brute force variant, and by a factor of 2.4–17.1 compared to the depth bounds. In the shadow volume pass, our algorithm processed 5.3–22.9 times fewer pixels than brute force, and 2.8–11.5 times fewer than depth bounds.

We argue that it is the number of processed pixels that ultimately limits the performance. The bandwidth to external memory is a smaller problem, because less than three bytes were transferred per pixel for Knights, Powerplant, and Amphitheater. Also, it might be possible to reduce the bandwidth further by employing compression techniques to the stencil buffer. For Powerplant, our largest scene, as many as 2.2 billion pixels were processed every frame by the brute force algorithm. The depth bounds reduced the number of pixels to 943 million, and our algorithm provided a further reduction to 99 million. In an additional test, the brute force variant ran at less than one frame per second on a 3GHz Pentium 4 with an ATI Radeon 9800XT graphics card. When the shadow volume rasterization was omitted, the frame rate jumped to over 50. Thus the overall performance of the application was clearly bounded by the processing of shadow volumes. Assuming a 20-fold speedup in the shadow volume rasterization, a hardware implementation of our algorithm should boost the overall frame rate from 1 to around 15.

It may seem surprising that the Z-buffer bandwidth is also reduced by our algorithm. This results from skipping the processing of tiles that are already in shadow due to previous shadow volumes. In the Powerplant scene, a further reduction of bandwidth usage of 15MB (5%) was obtained when the cache for the PSV-buffers was doubled from 2kB to 4kB.

It is worth emphasizing that the performance ratio between our algorithm and depth bounds gets even larger with higher resolutions, because the rasterization work done by our algorithm is primarily affected by the length of the shadow boundary. In contrast, with depth bounds at least the whole shadow area needs to be rasterized accurately. When resolution increases, the length of the shadow boundary typically grows much slower than the covered area. At 2048 × 1536 resolution, our algorithm performed an additional 20–30% better than depth bounds.

Discussion The performance of our algorithm could be further improved by combining it with depth bounds testing, which offers a complementary and inexpensive way to discard tiles prior to Stage 1. In applications that always extrude the shadow volumes to infinity instead of the attenuation range the performance of depth bounds testing could be further improved by limiting the bounds with bounding volumes of light sources and with scene geometry. Under such circumstances, accurately computed depth bounds would allow skipping boundary tiles that are not affected by the light source, and thus also optimize the per-pixel operations in Stage 2.

The performance improvements are a result of hierarchical processing of non-boundary tiles. The worst case input for our algorithm would be a very complex object, such as a tree that could at least theoretically cast shadows that have a boundary inside every tile. In such cases, hierarchical processing cannot be exploited due to the complete lack of coherence in shadowed regions.

Highly tessellated shadow casters, such as the characters in the Knights scene, result in long and narrow shadow volume triangles that may not cover even a single tile entirely. That does not reduce the performance of our algorithm because we consider the shadow volume as a whole, and the size of individual triangles is not important.

7. Conclusions and Future Work

We have presented a new algorithm for accelerated rendering of shadow volumes. For a single-pass implementation using a delay stream, the benchmarks indicate that our shadow volume culling algorithm processes 2.8–11.5 times fewer pixels, and performs between 2.4 and 17.1 times fewer accesses to external memory than previous methods. Thus, we argue that the main critique against shadow volumes is eliminated by our method.

In future work, we would like to make an accurate comparison of computational and bandwidth requirements of our hierarchical shadow volume algorithm and the shadow mapping algorithm. A performance evaluation with different tile sizes is also an interesting line of future work. Furthermore, it might be fruitful from a performance viewpoint to study in which order the shadow volumes should be submitted.

Acknowledgments

Simple scene courtesy of Tom McReynolds, SGI. Powerplant courtesy of the Walkthrough Group at the University of North Carolina at Chapel Hill. Amphitheater courtesy of Sam Howell, used with permission of Morgan McGuire. The Tick model used in Amphitheater was created by Carl Schell (carl@cschell.com). The Perelith knight model courtesy of James Green (www.perilith.com, james@perilith.com). Thanks to Jonas Svensson and Ulf Borgenstam for providing the Knights demo, Karl Schultz for help with MESA, and Ville Miettinen, Jacob Ström, Eric Haines, Ulf Assarsson, Lauri Savioja and the 3DR group at Helsinki University of Technology for helpful comments. Timo Aila was partially supported by the National Technology Agency of Finland, Bitboys, Hybrid Graphics, Nokia and Remedy Entertainment.

Appendix A: Pseudo Code

Stage 1 (Z-fail)

```
for each tile t overlapped by the triangle:

    if (triangle intersects 3D bounding box of t)
        t.PSV.boundary = true;

    else if (triangle behind 3D bounding box of t &&
             t's sample point inside triangle)
        if (triangle front-facing)
            t.PSV.stencil = t.PSV.stencil - 1;
        else
            t.PSV.stencil = t.PSV.stencil + 1;
```

Stage 2 (Optimized Architecture)

```
for each tile t overlapped by the triangle:

    if (t.PSV.first == true)
        t.PSV.first = false;
        if (t.Smin > Sclear)    // in shadow before this SV
            t.FS = true;

    if (t.FS == true)           // already in shadow
        continue;

    if (t.PSV.boundary == true)
        rasterizeToStencilBuffer(); // updates Smin/Smax too

    else if (t.PSV.stencil > Sclear)  // full shadow
        t.Smin = t.Smax = t.PSV.stencil;
        t.FS = true;
```

References

[AAM03] Assarsson U., Akenine-Möller T.: A Geometry-based Soft Shadow Volume Algorithm using Graphics Hardware. *ACM Transactions on Graphics, 22*, 3 (2003), 511–520.

[ADMAM03] Assarsson U., Dougherty M., Mounier M., Akenine-Möller T.: An Optimized Soft Shadow Volume Algorithm with Real-Time Performance. In *Graphics Hardware* (July 2003), ACM SIGGRAPH/Eurographics, pp. 33–40.

[AMN03] Aila T., Miettinen V., Nordlund P.: Delay Streams for Graphics Hardware. *ACM Transactions on Graphics, 22*, 3 (2003), 792–800.

[AMS03] Akenine-Möller T., Ström J.: Graphics for the Masses: A Hardware Rasterization Architecture for Mobile Phones. *ACM Transactions on Graphics, 22*, 3 (2003), 801–808.

[CD04] Chan E., Durand F.: An Efficient Hybrid Shadow Rendering Algorithm. In *Proceedings of the*

Eurographics Symposium on Rendering (2004), Eurographics Association.

[Cro77] CROW F.: Shadow Algorithms for Computer Graphics. In *Computer Graphics (Proceedings of ACM SIGGRAPH 77)* (July 1977), ACM, pp. 242–248.

[EK02] EVERITT C., KILGARD M.: Practical and Robust Stenciled Shadow Volumes for Hardware-Accelerated Rendering. *http://developer.nvidia.com/* (2002).

[GLY*03] GOVINDARAJU N. K., LLOYD B., YOON S.-E., SUD A., MANOCHA D.: Interactive Shadow Generation in Complex Environments. *ACM Transactions on Graphics, 22*, 3 (2003), 501–510.

[Hei91] HEIDMANN T.: Real Shadows, Real Time. *Iris Universe*, 18 (November 1991), 28–31.

[HM01] HAINES E., MÖLLER T.: Real-Time Shadows. In *Proceeding of Game Developers Conference* (March 2001), pp. 335–352.

[Len02] LENGYEL E.: The Mechanics of Robust Stencil Shadows. *http://www.gamasutra.com/* (October 2002).

[LWGM04] LLOYD B., WENDT J., GOVINDARAJU N., MANOCHA D.: CC Shadow Volumes. In *Proceedings of the Eurographics Symposium on Rendering* (2004), Eurographics Association.

[MBGS01] MEISSNER M., BARTZ D., GÜNTHER R., STRASSER W.: Visibility Driven Rasterization. *Computer Graphics Forum, 20*, 4 (2001), 283–294.

[McC00] MCCOOL M. D.: Shadow Volume Reconstruction from Depth Maps. *ACM Transactions on Graphics, 19*, 1 (2000), 1–26.

[MHE*03] MCGUIRE M., HUGUES J. F., EGAN K. T., KILGARD M., EVERITT C.: *Fast, Practical and Robust Shadows*. Tech. Rep. CS03-19, Brown University, October 2003.

[Mor00] MOREIN S.: ATI Radeon HyperZ Technology. In *Workshop on Graphics Hardware, Hot3D Proceedings* (August 2000), ACM SIGGRAPH/Eurographics.

[NVI03] NVIDIA: *NVIDIA GeForceFX 5900 GPUs: UltraShadow Technology*. Tech. rep., http://www.nvidia.com, 2003.

[SCH03] SEN P., CAMMARANO M., HANRAHAN P.: Shadow Silhouette Maps. *ACM Transactions on Graphics, 22*, 3 (2003), 521–526.

[SHSG01] SANDER P. V., HOPPE H., SNYDER J., GORTLER S. J.: Discontinuity Edge Overdraw. In *Symposium on Interactive 3D Graphics* (March 2001), ACM SIGGRAPH, pp. 167–174.

[Wil78] WILLIAMS L.: Casting Curved Shadows on Curved Surfaces. In *Computer Graphics (Proceedings of ACM SIGGRAPH 78)* (August 1978), ACM, pp. 270–274.

[WPF90] WOO A., POULIN P., FOURNIER A.: A Survey of Shadow Algorithms. *IEEE Computer Graphics and Applications 10*, 6 (November 1990), 13–32.

Graphics Hardware (2004)
T. Akenine-Möller, M. McCool (Editors)

Squeeze: Numerical-Precision-Optimized Volume Rendering

Ingmar Bitter
Clinical Image Processing Services
Department of Radiology
National Institutes of Health
IBitter@nih.gov

Neophytos Neophytou, Klaus Mueller, Arie E. Kaufman
Computer Science
Center for Visual Computing
State University of New York at Stony Brook
{nneophyt|mueller|ari}@cs.sunysb.edu

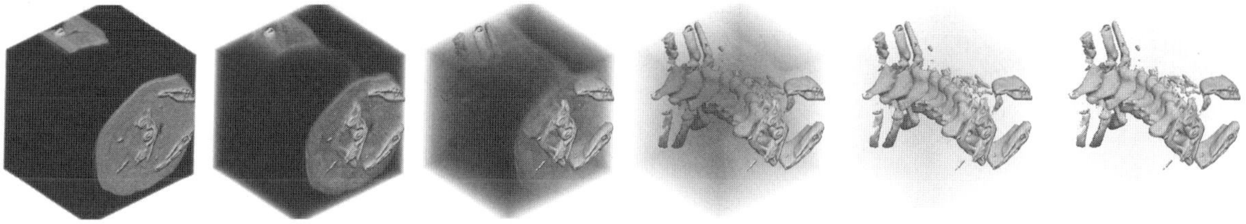

Abstract
This paper discusses how to squeeze volume rendering into as few bits per operation as possible while still retaining excellent image quality. For each of the typical volume rendering pipeline stages in texture map volume rendering, ray casting and splatting we provide a quantitative analysis of the theoretical and practical limits for the required bit precision for computation and storage. Applying this analysis to any volume rendering implementation can balance the internal precisions based on the desired final output precision and can result in significant speedups and reduced memory footprint.

Categories and Subject Descriptors (according to ACM CCS): I.3.3 [Computer Graphics]: Picture/Image Generation E.2 [Data]: Data Storage Representations

1. Introduction

Volume rendering has undergone a tremendous evolutionary process in the past decade and it has found a wide range of application domains in which it is routinely being used. The most obvious are the visualization of medical and scientific datasets, where the input data are most often inherently volumetric. However, we have also seen applications using volumetric sculpting to interactively generate volumetric data using volumetric deformation to modify the data on the fly. Many advances in software rendering algorithms [LL94, Lev88, MJC02, Wes90], data structures [IL95, PSL*98, WV92], and custom hardware [MKW*02, PHK*99] have been made, and the recent revolution in commodity graphics hardware has helped volume rendering to gain speedups on the order of 1-2 magnitudes [EKE01, RSEB*00]. What has remained quite stable over the years is the volume rendering pipeline itself. Although we have seen a shift from pre-classified pipelines to post-classified ones (shown in Figure 1), this only requires a re-ordering of the pipeline stages to facilitate classification and shading after interpolation [MMC99]. Due to the massive amount of data that is involved in volume rendering, caching has received a fair bit of atten-

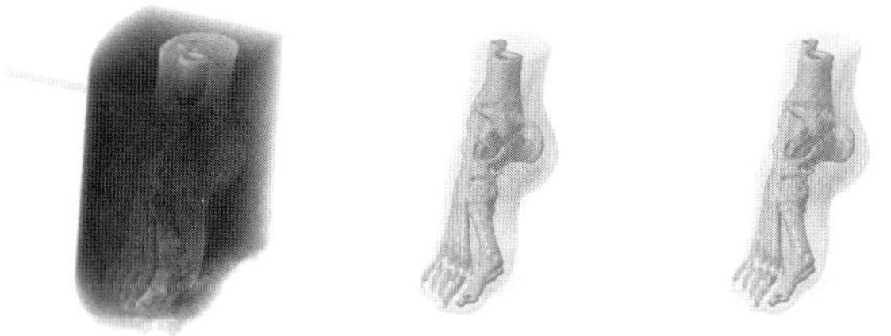

Figure 1: *Thin soft tissue using* $\alpha^{0.8b}$, $\sqrt{\alpha}^{0.8b}$, *and* $\alpha^{0.16b}$.

tion [Kni00, LCCK02, PSL*98]. All this resulted in a great deal of understanding on how to order the data items as they flow across the pipeline. However, what about the size of the data items? This is just as important because the size of the data items influences (i) the computational effort, (ii) the amount of data items that can be squeezed into the caches and registers, and (iii) the amount of data bandwidth that can be attained across both networks and buses. In this paper, we have attempted to address this important issue by analyzing the required data item precision at every stage of the volume rendering pipeline, and in this way to come up with the optimal data item size. The importance of maintaining sufficient precision was also observed in earlier work [MHB*00] when comparing several popular volume rendering approaches. In that research it was noticed that structures of low opacities could not be visualized well when using texture mapping hardware with limited bit-precision, especially when the pre-classified pipeline was applied in conjunction with opacity-weighted color interpolation [WMG98].

In our discussion we have assumed that the precision is ultimately determined by the final stage, that is, the display on a computer monitor or projection wall, which is in most cases 8 bits per color channel. Although input and output media with higher precision do exist and our discussion readily generalizes to those, we feel that 8 bits will be a hard limit for a while because of the same limit within the DVI standard used by most LCD screens and the absence of any commodity displays with higher color resolution. We have also concentrated on two major volume rendering algorithms, raycasting and splatting. They can be implemented in four distinct ways. (i) First, one can write a software volume renderer that uses general purpose CPUs to compute the results of each raycast or splat volume rendering pipeline stage. (ii) Second, one can implement the same raycast or splat volume rendering pipeline stages using the same ray patterns and the same math at 32-bit precision using custom written shader programs for the new nVIDIA FX and ATI Radeon 9700/9800 chips. Hence, the rendering pipeline analysis is the same for this approach, only that any result exceeding 32-bit precision can not be satisfied in the implementation. In addition, shader programs can use textures of varying bit depth and our analysis helps to decide which one to use. (iii) Third, one can use standard OpenGL or DirectX to implement texture map volume rendering that combines slices of 2D or 3D textures to form the volume rendered image. Mapping OpenGL or DirectX texture map functions to the splatting math is difficult and we are not aware of any published implementations. For raycasting a perfect match does not exist as well, but a significant subset is implementable. We point out where the general pipeline math can not be cast into standard OpenGL or DirectX calls. Texture map raycasting uses the same rendering pipeline stages, but is easier to implement with classification before interpolation. In order to accomodate this approach we analysed each pipeline stage seperately. The order inversion of these two stages actually have no effect on the outcome of the required precision. (iv) The forth approach to implementing raycasting or splatting is to build custom hardware. Here the sky is the limit, but hardware resources need to be carefully balanced to achieve maximum performance and our rendering pipeline analysis gives valuable pointers of how much precision to invest in each stage.

Our paper is structured as follows. First, Section 2 will discuss related work, while Section 3 will present some theory on fixed-point arithmetic in the context of our special volume rendering scenario. Then, Section 4, will present our detailed precision analysis, addressing both theoretical and practical issues. Section 5 will present some results, and Section 6 will draw final conclusions.

2. Related Work

The VolumePro volume rendering chip [PHK*99] uses 12 bits throughout the pipeline, mainly motivated by the need to cater to the medical imaging industry that uses 12-bit data words. The Vizard II board [MKW*02] uses a 16-bit fixed-point format in the compositing stage to achieve good image quality. In the color lookup tables, it allocates more bits (16) for opacities than for colors (8) to resolve low opacities when oversampling the rays. The PentiumIII based UltraVis system [Kni00] achieves high frame rates through extremely cache-friendly data layout and access. UltraVis uses 8-bit data as well as 16-bit fixed-point color and transluceny buffers in the compositing stage. Finally, two major commodity graphics board manufacturers nVIDIA and ATI have recently made available 32-bit floating-point arithmetic within their texture shaders, which brings plenty of precision, but requires non-standard vendor specific shader programs.

3. Fixed-Point Arithmetic for Volume Rendering

In order to conduct our precision analysis within a common framework, we will be using fixed-point notation. This is convenient since for volume rendering the range of useful results is generally contained in the tight interval of [0.0, 1.0]. This restricted interval has a number of important, as well as peculiar consequences. Note, that in any fixed-point representation $I.Fb$ with I bits for the integer part and F bits for the fractional part the number 1.0 is represented as $I = 1$ and $F = 0$ or 1.0b. With this representation multiplying 1.0b with itself is achieved through multiplication of the integer variable with itself and a right shift by F bits, which results again in 1.0b. However, in OpenGL the range of α can be expressed as [0.0f,1.0f] as well as [0,255] which equates 1.0f with 255. If we use the [0,255] representation and try to compute 1^2 with one multiply and one shift we get $255^2 >> 8 = 254$. Due to truncation effects each successive multiplication with 255 will further reduce the result by 1. As we will see in the section about compositing 4.1 that we need to successively multiply total transparency (1.0). Hence, with the [0,255] representation of the unit range any attempt to composite complete transparency for 255 or more samples will counterintuitively result in total opacity. Consequently, we need more computational effort to use the [0,255] representation. One multiply, two adds and two shifts properly compute 1^2 [Bli95]:

$$\begin{aligned} tmp &= a^{[0,255]} * b^{[0,255]} + 128; \\ result &= (tmp + (tmp >> 8)) >> 8; \end{aligned} \quad (1)$$

However, a proper $I.Fb$ fixed-point format can be used to compute the transparency compositing using only one multiply and one shift. With this approach the eliminated second shift and adds save half the computation time compared to the [0,255] representation. Therefore, the one saved bit in the [0,255] representation is not worth the computational cost, in particular in light of the fact that 8 bits are by far insufficient for the general case, as we shall see below. On a Pentium 4 both operations together actually take longer than a single floating-point multiplication. Hence, for those CPUs it only pays to apply the fixed-point arithmetic in practice if the need for multiplications is mixed with additions, in which case a single integer add can be used which is much faster than a floating-point add.

4. Precision Analysis

Figure 2 illustrates the post-classification rendering pipeline stages for raycasting and splatting (specifically image-aligned sheet-buffered splatting [MSHC99]). The structure of both algorithms is very similar and so are most of their numerical precision requirements. Eventually both approaches result in an image that is usually displayed on a computer monitor. We focus here on mainstream PCs with 8 bits per color channel for the final display. These 8 bits per color channel are the crucial limitation for the achievable precision. We optimize all prior steps to have just enough precision to guarantee that the final image precision is not harmed. To propagate the limitations through the rendering pipeline, we analyze the pipeline stages in reverse order.

4.1. Compositing

The last step in the rendering pipeline is the compositing of the shaded samples yielding the final pixel value. For realistic-looking images $RGB\alpha$-tuples affecting the same pixel are interpreted as samples along a light ray into this pixel. Further, the light attenuation when traversing semi-transparent colored material is simulated [KV84, Kru91] using the *OVER* operator (see Equation 2) to ***composite*** consecutive $RGB\alpha$-tuples along the ray in sorted order. This blends the tuples with the proper partial occlusion. The computations have to be applied for each of the red, green and blue color channels, with any one of them denoted as C_λ.

$$\begin{aligned} {}^{OVER}\alpha C_\lambda &= C_{\lambda front} * \alpha_{front} + C_{\lambda back} * \alpha_{back} * (1 - \alpha_{front}) \\ {}^{OVER}\alpha &= 1 - (1 - \alpha_{front}) * (1 - \alpha_{back}) \end{aligned} \quad (2)$$

Compositing all samples contributing to the same pixel with the *OVER* operator in front to back order is called ***front to back compositing (F2B)***. It can be done through incremental compositing of color $C_{\lambda_{back}}$ and α_{back} into a compositing buffer (CB):

$$\begin{aligned} {}^{F2B}\alpha C_{\lambda CB} &+= C_{\lambda back} * \alpha_{CB} * (1 - \alpha_{CB}) \\ {}^{F2B}\alpha_{CB} &= 1 - (1 - \alpha_{CB}) * (1 - \alpha_{back}) \end{aligned} \quad (3)$$

(the += operator is borrowed from the C/C++ language). It is most efficient to incrementally composite α-pre-multiplied colors $\widehat{C}_\lambda = \alpha C_\lambda$ and transparency $T = 1 - \alpha$, thus, compositing $(\alpha R, \alpha G, \alpha B, T)$-tuples:

$$\begin{aligned} {}^{F2B}\widehat{C}_{\lambda CB} &+= \widehat{C}_{\lambda back} * T_{CB} \\ T_{CB} &*= T_{back} \end{aligned} \quad (4)$$

Blending with the *OVER* operator in the reverse direction is called ***back to front compositing (B2F)*** and results in the same final color:

$$\begin{aligned} {}^{B2F}\alpha C_{\lambda CB} &*= 1 - \alpha_{front}; \quad {}^{B2F}\alpha C_{\lambda CB} += C_{\lambda front} * \alpha_{front} \\ {}^{B2F}\alpha_{CB} &= 1 - (1 - \alpha_{CB}) * (1 - \alpha_{front}) \end{aligned} \quad (5)$$

Again, it is more efficient to composite $\widehat{C}_\lambda$ and T.

$$\begin{aligned} {}^{B2F}\widehat{C}_{\lambda CB} &*= T_{front} += \widehat{C}_{\lambda front} \\ T_{CB} &*= T_{front} \end{aligned} \quad (6)$$

Compositing all contributions along a ray yields the final transparency T_{CB}, which is a material property that is independent of the sampling rate. Hence, the transparency given

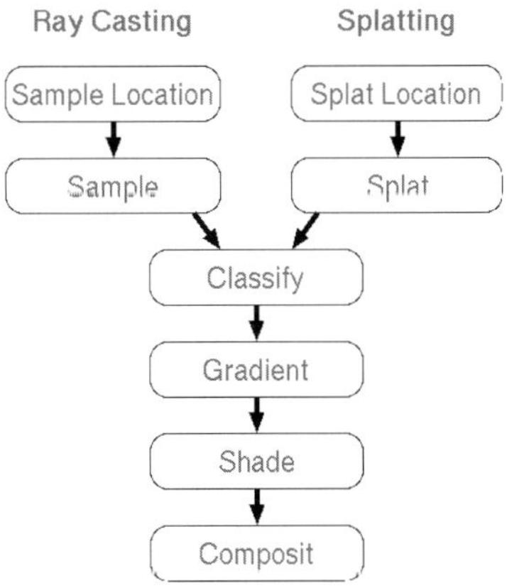

Figure 2: *The rendering pipeline stages for raycasting and splatting.*

to each sample has to be adjusted accordingly if the sampling rate is altered. For a homogeneous semi-transparent gel and a sampling rate of one sample per voxel and N samples along a ray, the final composited transparency becomes $T_{CB_N} = T^N$ (using Equation 6). If we now sample K times along the same ray, the sampling rate changes to $r = K/N$ and the new final transparency is $T_{CB_N} = \tilde{T}^K$. As the material has not changed, both expressions have to be equivalent:

$$\begin{aligned} \tilde{T}^K &= T^N \\ \tilde{T} &= T^{N/K} = T^r \\ \tilde{\alpha} &= 1-(1-\alpha)^r \end{aligned} \qquad (7)$$

Adjusting α according to Equation 7 is called ***alpha correction*** [LCN98].

In a slice-order rendering algorithm, such as splatting or texture map volume rendering or Cube-4 style raycasting [PK96], a complete slice of the intermediate results is stored in the compositing buffer which requires a significant amount of memory. A 512^2 $RGB\alpha$-image in floating-point format uses 4MB, which is well beyond the on-chip cache sizes of special-purpose volume rendering chips. It is also well beyond the sizes of general purpose processors level-1 and level-2 caches. Hence, reducing the compositing buffer to the numerically required minimum will improve the cache hit ratio.

4.1.1. Theoretical Analysis

Multiplying two numbers in fixed-point format with f_1 and f_2 fractional bits each results in a new number with $f_1 + f_2$ fractional bits. During B2F and F2B compositing each sample requires a multiplication for color as well as for opacity or transparency. Thus, compositing of N samples requires Nf fractional bits for an absolute correct result. However, when constrained to a frame buffer with 8-bit precision, the compositing buffer precision may be reduced.

Compositing color $\widehat{C}_{\lambda CB}$ through Equation 4 or 6 accumulates the sample color contributions. With $N = 2^p$ samples along a ray, small contributions sum up such that the same low significance bit in each summand can influence a higher significance bit p positions further to the left in the final sum. The exponent p depends on the volume size and sampling rate, and it ranges from 8 for 256^3 volumes and one sample per voxel to 11 for $512^2 \times 2048$ volumes. Hence, to allow 8-bit precision of the final color, $8+p$ bits are needed to account for this accumuation of small contributions. Note, that these small contributions may come from a saturated color being multiplied with a small opacity (or high transparency). Hence, opacity should be represented with with $8+p$ fractional bits as all of those can potentially affect the final composited result. Further, the final display of the rendered image is assumed to be on a 24-bit (R,G,B) display device which requires a conversion of $\widehat{C}_{\lambda CB} = \alpha C_{\lambda CB}$ to just $C_{\lambda CB}$. This conversion needs a division by α. If only the least significant bit of α is non-zero, then during the division the bits in $\alpha C_{\lambda CB}$ are shifted to the left by $8+p$ positions. Thus, over all, to allow 8-bit precision of the final color, $2(8+p)$ bits are needed to composite $\widehat{C}_{\lambda CB}$ which adds up to 32-38 bits.

Compositing using F2B (Equation 4) or B2F (Equation 6) both compute the transparency T through successive multiplications. If T is represented with the same precision as α using $t = 8+p$ fractional bits then T it is a number in $1.t$ format and is always in the value interval [0.0, 1.0]. Note, that during multiplication with any other number in the interval [0,1], none of the bits after the tth have any effect on the first t fractional bits of the multiplication result. Hence, reducing the transparency compositing buffer precision to the same $8+p$ bits as each sample's transparency is sufficient, which for our volume sizes and sampling rates equates to 16-19 bits. Note, that this is only half for the precision needed for $\widehat{C}_{\lambda CB}$.

Compositing $RGB\alpha$ through Equation 3 or 5 adds more operations, but the number of multiplications and the number of sample contributions to sum up do not change. Hence, the precision analysis of $(\alpha R, \alpha G, \alpha B, T)$ also applies to $RGB\alpha$.

At a sampling rate r, applying alpha correction adjusts the transparency T to $\tilde{T} = T^r$. For sub-sampling $(r < 1)$ there are no precision limitations, but for super-sampling $(r > 1)$ the value of any $\tilde{T} \in (0,1)$ is less than that of T. The smallest change is needed for values close to zero and the smallest value of T at f fractional bits is 2^{-f}. Even for this value $0 < \tilde{T} < T$ still has to be true.

$$\begin{array}{ccccccccc} 0 & < & \tilde{T} & = & T^r & < & T & = & T_{min} \\ 0 & < & \tilde{T} & = & \left(2^{-f}\right)^r & < & T & = & 2^{-f} \\ 0 & < & \tilde{T} & = & 2^{-fr} & < & T & = & 2^{-f} \end{array} \qquad (8)$$

Equation 8 shows that the number of bits needed to represent $\tilde{T}$ is by a factor r larger than the bits needed for T. Consequently, the assumed $8+p$ bits in the previous paragraphs need to be increased to $r(8+p)$.

The overall compositing buffer precision requirements are $r2(8+p)$ for $\widehat{C}_{\lambda CB}$ and $r(8+p)$ for T_{CB} and α_{CB}. This adds up to 1*2*(8+8) = 32 bits for T_{CB} and 1*(8+8) = 16 bits for α_{CB} for 256^3 volumes and no super-sampling. For $512^2 \times 2048$ volumes and 16 samples per voxel we get 16*2*(8+11) = 608 bits for T_{CB} and 304 bits for α_{CB}.

4.1.2. Practical Considerations

If we ignore the precision requirements due to alpha correction, we set r to 1 and just slightly modify the transparency of some structures without changing the overall impression. Fortunately, the largest error occurs at low transparencies, affecting samples that do not have much of a chance to accumulate the error. This is because low transparency is synonymous to most opaque and any contributions of samples behind a nearly opaque sample are usually not noticeable. Incorrectly snapping 99% opaque to 100% opaque is very acceptable. With this approximation the requirement for $\widehat{C}_{\lambda CB}$ becomes $2(8+p)$ bits and for T_{CB} or α_{CB} $8+p$ bits.

4.1.2.1. Opaque Iso-surface Visualization If texture map volume rendering is supposed to display only one or more iso-surfaces and if these are to be rendered as opaque surfaces, then the compositing buffer precision can be even more reduced. Storing the colors as $C_{\lambda CB}$ requires a division by α for each sample, but reduces the bit precision requirement to $8+p$ bits. Additionally, p can be reduced to 3, as usually no more than 8 non-zero samples are needed to reach opaque surface appearance. If we further assume that the colors assigned by the transfer function are very bright and can deal with a 3% inaccuracy (≥8/256), then we can delete 3 more bits from the input color resulting in $5+p = 5+3 = 8$ bits for $C_{\lambda CB}$. *Hence, with opaque iso-surface oriented visualization configurations using 8 bits per channel are sufficient.* Note, that this restricted scenario allows texture map volume rendering to perform the compositing in the frame buffer with sufficient precision.

4.1.2.2. Fog Visualization Practical scenarios in which more bits are needed occur when an object is partially hidden in fog or bones are partially hidden behind nearly completely transparent soft tissue. Here, we cannot ignore the accumulation effect of small contributions building up over a long stretch of low opacity samples. If the volume is covered in fog and a structure whose iso-surface is supposed to be displayed is centered in the volume, each ray reaching this structure first traverses fog for about half the volume. This adds up to 1024 samples for a 512^3 volume and a sampling rate of 4 samples per voxel. In this scenario we have two special cases due to the high transparency values that need to be successively composited to T_{CB} according to Equation 4.

The first special case is concerned with the successively compositing of completely transparent material, and we already showed in Section 3 that we need to represent the transparency T as $1.Fb$ format number to maximize computation efficiency and ensure that $1^2 = 1$. The second special case is the soft tissue or fog that we do want to see, but only highly transparent such that the opaque structure or bone within is still fully visible. Using a fixed-point format $1.7b$ with only 7 bits for the fractional part F, $\frac{2^7-1}{2^7} = \frac{127}{128}$ is the smallest partial transparency selectable. With this setup, each fog sample reduces the ray transparency by $\frac{1}{128}$, reaching full opacity after 63 samples. Therefore, the structure would be completely hidden by the fog. Increasing the fractional precision to 8, 10, 12, 14, 15, and 16 bits results in 1024 samples of thin fog accumulating to 100%, 100%, 25%, 6.25%, 3.08%, and 1.7% opacity, respectively. Note, that here the color compositing buffer channels need at least the same precision as the transparency channel, but the per sample color does not need such high precision as it is used only once.

The teaser images on the title page illustrate the effect of the compositing buffer bit depth $b = 8, 10, 12, 14, 15, 16$ when rendering a human spine from a 512^3 dataset with a sampling rate of 2 and with all non-bone tissues and the background set to $\alpha = \frac{1}{2^b}$. This enables only the least significant bit and completely encloses the spine in this least-significant-bit-fog. It is clear that for this type of visualization 8 bits or 10 bits are not enough. 12 bits may be sufficient, but adjusting the fog density will only be possible in very noticeable discrete steps. 14 bits is sufficient for smooth fog adjustments. 15 bits would even allow increasing the sampling rate from 2 to 4 and still result in an image similar to the 14-bit fog image. *Hence, for general visualization settings, the compositing buffer should have 15 bits per color channel. A hardware implementation geared towards general volume visualization and 512^3 maximum volume dimension may reduce the compositing buffer to 14 bits per color channel.*

Placing the preferred fixed-point bit precision 1.15b as "exponent", the preferred compositing equation is:.

$$\begin{aligned} {}^{F2B}\widehat{C}_{\lambda CB}^{1.15b} &\mathrel{+}= \widehat{C}_{\lambda back}^{1.15b} * T_{CB}^{1.15b} \\ T_{CB}^{1.15b} &\mathrel{*}= T_{back}^{1.15b} \end{aligned} \tag{9}$$

4.2. Shading

The color C_λ used in compositing results from evaluating the Phong illumination:

$$\begin{aligned} {}^{Phong}C_\lambda &= k_{ambient} O_\lambda I_\lambda \\ &+ k_{diffuse} O_\lambda \sum_{i=1}^{N_{lights}} I_{\lambda_i} \left(\vec{N} \bullet \vec{L}_i\right) \\ &+ k_{specular} \sum_{i=1}^{N_{lights}} I_{\lambda_i} \left(\vec{R}_V \bullet \vec{L}_i\right)^r \end{aligned} \tag{10}$$

For OpenGL and DirectX texture map volume rendering the specular term must be dropped because of the inability to compute the reflectance vector. However, it could be computed inside a custom shader program.

4.2.1. Theoretical Analysis

In Equation 10 the object color O_λ is usually an 8-bit value representing the range [0,1] read from a lookup table. Further, each I_{λ_i} is in the range [0,1] and the vectors are supposed to be normalized such that their dot product is in the range [-1,1]. However, a negative dot product signifies that the local surface element is facing away from the eye point such that its contribution should better be culled. Hence, all negative dot products are clamped to zero, which reduces the dot product result range to [0,1]. In addition, the sums $\sum_{i=1}^{N_{lights}}$ are clamped to the [0,1] range as well, and finally one can assume that $k_{ambient} + k_{diffuse} + k_{specular} = 1$ as those constants represent the weighting of the intensity contributions due to ambient, diffuse and specular light. Consequently, the overall sum will be in the range [0,1].

4.2.2. Practical Considerations

Knowing from Section 4.1.2 that the computed color needs 15 fractional bits and that the value range is [0,1], a $1.15b$ format to represent ${}^{Phong}C_\lambda$ is sufficient. However, each of the multiplications needs additional bits. These bits can either come from using standard floating-point numbers or through the use of an extended fixed-point format. If for each product one factor is in the range [0,1] and the other in the range [0,1), then a balanced fixed-point format such as 16.16b will not overflow during the integer multiplication part. And if after each multiplication the result is shifted right by 16 bits all consecutive multiplications with more values in the [0,1] range do not overflow as well. If any of those factors is in the range (0,1) but still truncates the multiplication result to zero because of limited precision, we only incur a neglible error as the number of lights is usually small. Given eight lights the sum $\sum_{i=1}^{N_{lights}}$ could bring back eight contributions that are one bit below the least significant to the second to last position of the valid range. On the 16.16b format this is the difference between 0.003% intensity and 0% intensity, which is a very acceptable error. The only operation during the Phong equation evaluation that can not easily be replaced by integer math is the exponentiation needed for the specular light contribution. Knittel [Kni00] proposed to replace the exponential function with an inverted parabola that has a very similar shape and very similar area under the curve. Unfortunately, this approach has a discontinuous first derivative at the location where the contribution approaches zero. This discontinuity in the change of brightness is very visible to the human eye (see Figure 3). A better method is to restrict the exponent to powers of two and to compute the power through successive multiplications of intermediate results. If each multiplication truncates the results to 16 fractional bits then a value raised to the power of $2^6 = 64$ will have a maximum error of $1 - (1 - \frac{1}{2^{17}})^6 = 0.0046\%$ which results in much smoother perceived changes in brightness (see Figure 3). *Hence, for the use of fixed-point math the Phong illumination equation can be evaluated using the 16.16b fixed-point format:*

$$\begin{aligned} {}^{Phong}C_\lambda^{16.16b} &= k_{ambient}^{16.16b} O_\lambda^{0.8b} I_\lambda^{16.16b} \\ &+ k_{diffuse}^{16.16b} O_\lambda^{0.8b} \sum_{i=1}^{N_{lights}} I_{\lambda_i}^{16.16b} \left(\vec{N}^{16.16b} \bullet \vec{L}_i^{16.16b}\right) \\ &+ k_{specular}^{16.16b} \sum_{i=1}^{N_{lights}} I_{\lambda_i}^{16.16b} \left(\vec{R}_V^{16.16b} \bullet \vec{L}_i^{16.16b}\right)^{2^n} \end{aligned} \tag{11}$$

4.3. Gradients

Our analysis is based on the most commonly used gradient estimator, the Central Difference Gradient $G_{x,y,z}$, with $G_x = \frac{1}{2}sample_{(x+1,y,z)} - \frac{1}{2}sample_{(x-1,y,z)}$, G_y and G_z analog.

4.3.1. Theoretical Analysis

The central difference gradient computation is straightforward. Given samples in a $I.Fb$ format the difference operator adds one bit of precision requirement to the integer part I and the division by 2 one bit to the precision requirement of the fractional part F. If the division by 2 is performed first, we do not even need the extra bit for the integer part. Hence, the result should be represented in a $I.(F+1)b$ format.

After normalization the gradient components are represented in 1.Fb format. This allows a discrete set of gradient directions with nearest neighbors having an angular difference of:

$$\sin\phi = \frac{1}{2^F}, \quad \sin\phi \approx \phi \;\Rightarrow\; \phi \approx \frac{1}{2^F}. \tag{12}$$

This difference ϕ has the most severe influence in the specular lighting computation. With an exponent of 64 the power of the original dot product $1^{64} = 1$ may shrink to $(1 - \frac{1}{2^F})^{64}$, which is a reduction by 22%, 6.1%, 1.6% and 0.4% if F is 8, 10, 12 or 14, respectively. Hence, for acceptable small differences in shading intensity, normalized gradients should

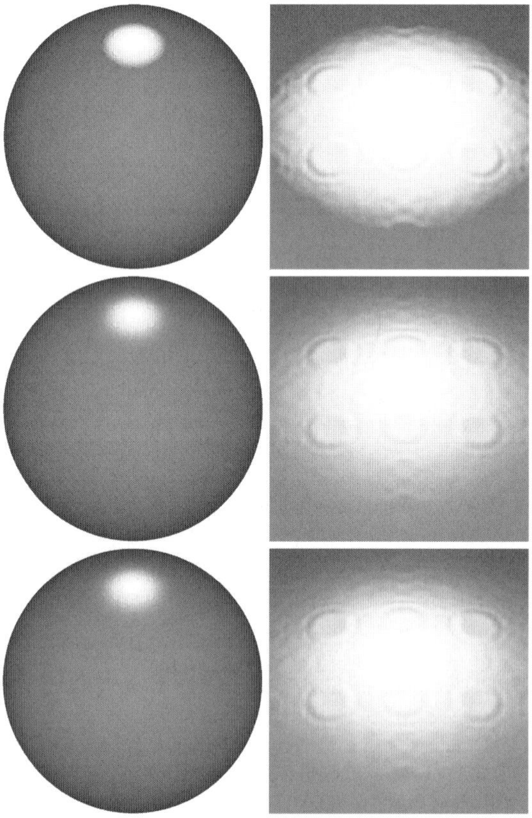

Figure 3: *A* 512^3 *sphere with Knittel's Phong approximation (top), with float precision and the pow() function of Phong (middle), and with fixed-point Phong (bottom).*

be accurate up to at least 12 fractional bits. Figure 4 displays spheres rendered with a specular highlight and gradient precision of 4, 6, 8, 10, 12 and 14 bits. We observe that for 4-bit precision there are significant quantization artifacts in form of rings for the diffuse shading and an almost nonexistent specular highlight. While 8-bit gradients are sufficient for the diffuse shading, the close-up images indicate that we need at least 12-bit gradients to obtain high-quality specular highlights.

4.3.2. Practical Considerations

For data from medical scanners the integer part of the samples often has 12 bits, while the fractional bits are set due to the sample interpolation computation. As shown in section 4.2 the normal used in the shading computation is expected to be normalized with components in the [0,1] range and in 1.15b format. If we perform this normalization the integer part of each component will be reduced from I to just 1 and there will be up to I bits added to the fractional part of each gradient vector component leaving only 4 interpolation bits to fill the remainder of the 16 significant fractional gradient vector bits. However, if the actual gradient magnitude is very small, then the normalization divides by a much smaller number and the fractional interpolation bits remain more significant. In the extreme all 16 bits can be influenced by the interpolation result.

In summary, both gradient computation and gradient normalization can be carried out using a 16.16b format.

This is a much harder requirement than can be satisfied by 8 bits per channel texture map implementations. Fortumnately, this requirement is mostly based on the specular reflection contribution which texture map volume rendering must ignore anyway. The remaining diffuse shading is sufficiently represented with 8 bits such that iso surface oriented texture map volume rendering still can be achieved without precision compromises.

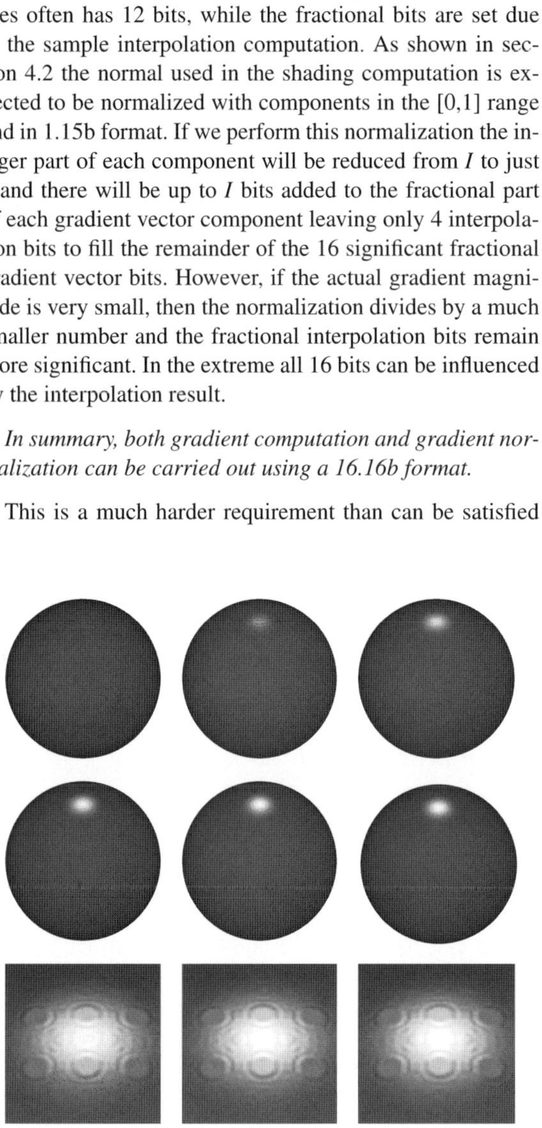

Figure 4: *First two rows: 512^3 spheres with gradient precision of 4, 6, 8, 10, 12, and 14 bits. Third row: close-ups for gradient precision of 10, 12, and 14 bits.*

4.4. Classification

Classification is the process of converting a voxel or sample into an $RGB\alpha$-tuple (the object color O_λ and the object opacity O_α). A ***color transfer function*** is used to map samples to colors. An ***opacity transfer function*** or ***α-transfer function*** is used to map samples to opacities α. It assigns high opacity to voxels and samples that should be visible and zero opacity to those that should be invisible.

4.4.1. Theoretical Analysis

As the transfer function may have high frequencies, the sampling rate used during rendering must be very high to avoid aliasing artifacts. The best way to achieve a high sampling rate without much computational cost is to use pre-integrated classification [EKE01, MGS02]. Beyond a data dependent sampling rate placing additional samples in between an existing pair of samples produces data values that are practically linearly distributed between the two corner values. Hence, one can restrict sampling during rendering to the corner samples and tabulate a very finely sampled integral for any pair of previous and current sample value.

4.4.2. Practical Considerations

With our focus on 12-bit input data a 1D transfer function table needs to have 4096 entries resulting in 32KB if 16 bits are used for each $RGB\alpha$ entry. The corresponding pre-integrated 2D table has 4096^2 entries, occupying 128MB. Because of this high storage requirement, precision is sometimes sacrificed and a 256^2 table of 8-bit $RGB\alpha$ is used instead, which occupies only 256KB. According to our previous analysis we have seen that we prefer to composite transparency and not opacity, and we have also seen that we need more bits in the transparency channel than in the color channels. Hence, a good compromise may be the $R^{0.8b}G^{0.8b}B^{0.8b}T^{1.15b}$ format. However, this results in 5 bytes per $RGB\alpha$-tuple in the lookup table, which is not very cache friendly. We have also shown that quantization errors for high transparency values have much larger effects than those for high opacities. Our preferred solution therefore uses only 8 bits, but distributes the values with increased concentration in the high transparency range. We achieve this by not storing T, but $\sqrt{\alpha}$ in the lookup table. Now, reading $\sqrt{\alpha}^{0.8b}$ from the table and then computing $T^{1.16b} = 1 - (\sqrt{\alpha})^2$ results in a 1.16b value for T. With this approach we use all 8 bits for the fractional part and, thus, doubling the number of possible values in the [0,1] range compared to using an 1.7b format for α. The reason to store a form of α and not T in the lookup table is that we can not distinguish between a 1.0f and $\frac{255^2}{256^2}$, which is a difference of 0.8%. This does not lead to perceivable image differences for these opacities, but would for high transparencies as shown earlier. Figure 1 depicts the differences in the thin fog scenario when using $\alpha^{0.8b}$, $\sqrt{\alpha}^{0.8b}$, and $\alpha^{0.16b}$.

Hence, the preferred lookup table entry has the format $R^{0.8b}G^{0.8b}B^{0.8b}\sqrt{\alpha}^{0.8b}$.

For texture map volume rendering this format is not an option and we have to resort to $R^{0.8b}G^{0.8b}B^{0.8b}\alpha^{0.8b}$, which, again, provides sufficient precision as long as only opaque iso-surfaces are being visualized.

4.5. Sample Interpolation / Splat Scan Conversion

To compute the value of a sample along a ray at an off-grid location, tri-linear interpolation is commonly used. Any tri-linear interpolation can be decomposed into seven linear interpolations. Similarly, when a splat is scan converted into a sheet buffer from its off grid splat center location, bilinear interpolation is used to find the splat weights at the pixel locations. Any bi-linear interpolation can be decomposed into three linear interpolations. Each linear interpolation involves four or three operations:

$$sample = voxel_0 * (1-w) + voxel_1 * w \quad (13)$$

$$= w * (voxel_1 - voxel_0) + voxel_0 \quad (14)$$

4.5.1. Practical Considerations

The most precision we require of samples is for their use in gradient computation. We capped the gradient precision earlier to the 1.12b format after normalization. The second use of samples is as input for transfer function lookup in which case 12-bit integers are needed. As voxels are represented as 12.0b format the weights need to be in 1.12b format to ensure that in the extreme case the two voxel values that are being interpolated differ only by one, there are still 12 bits used in the dynamic range of the interpolated results. Thus we have enough precision for both possible scenarios: gradient estimation in low-dynamic range neighborhoods and in high-dynamic range neigborhoods. In the former case, normalization by a small gradient magnitude will shift the lower bits back to the left, while in the latter case we need the bits for the actual gradient computation. Hence, by providing 12.12b precision for the samples we are able to deal with both scenarios and therefore will be able to pick out even small density variations accurately (which makes most sense in datasets with low noise levels).

The preferred linear interpolation formula is therefore:

$$sample^{12.12b} = w^{1.12b} * (voxel_1^{12.0b} - voxel_0^{12.0b}) + voxel_0^{12.0b} \quad (15)$$

Texture map volume rendering is again restricted to 8 bits and only appropriate for iso-surface rendering.

4.6. Sample / Splat Location

To follow a ray through the data volume sample locations along the ray are computed at off grid locations. This computation is usually done incrementally. Hence, the kth sample is at location $\vec{P}_0 + \sum \vec{V}_{inc}$. This involves k additions which can accumulate the error in the least significant bit of the $\vec{V}_{inc}$ and invalidate all of the last $\log k$ bits. Consequently, at four samples per voxel and a $512^2 \times 2048$ volume as many as 14 bits could be incorrect. The remaining bits of the location must be sufficient to represent an off-grid location that accurately interpolates the surrounding data values that uses 12 bits, as explained in the next section. And finally the non-fractional part has to have sufficient precision to enumerate all voxels, which requires up to 12 bits for $512^2 \times 2048$. Thus, theoretically, the only proper choice to compute ray sample locations is double precision.

4.6.1. Practical Considerations

In practice, the limited precision location representation in $I.Fb$ format has the effect that continuous rays are snapped to a discrete set of rays with an angular difference of

$$\sin\alpha = \frac{k\frac{1}{2^F}}{k}, \quad \sin\alpha \approx \alpha \;\Rightarrow\; \alpha \approx \frac{1}{2^F}. \quad (16)$$

Hence, with only 6 fractional bits those discrete rays differ by 0.9 degrees, with 12 bits they differ by 0.05 degrees, and with 16 bits by only 0.0009 degrees. For parallel projection all rays in a single frame have the same direction alleviating any precision influence on image quality. During rotations a limitation to 0.9 degree steps is not perceivable so 6 bits precision are sufficient. For perspective projection 0.9 degree steps between neighboring rays are not enough, but 0.05 are, as this allows to distribute 1024 rays across a 60 degree field of view.

Further, considering that we want to address locations in up to 2048 slices we need at least 11 bits for the integer part of the sample location. *Consequently, the minimum sample location precision is the 11.12b format.*

In splatting, we deal with one "ray" per voxel instead of the one ray per image pixel during raycasting. The splat-ray traverses from the voxel to the eye. For parallel projection again all rays are parallel and again 6 fractional bits are sufficient to allow for smooth rotation. In the perspective projection case we can have volumes as large as $512^3 \times 2048$ with as many as 2048 voxels projecting from a line perpendicular to the view direction towards the eye. This is a factor of two more than the ray density during ray casting. *Therefore, to include splatting, we need one additional fractional bit resulting in a minimum sample location format requirement of 11.13b.*

For efficient splatting it is best to transform all contributing voxels first and store their transformed locations in sheet buffer bins according to their distance from the image plane. In addition, a u-index into the pre-integration splat table is needed that is derived from the transformed z-coordinate. After that the z-coordinate can be dropped and only x and y have to be stored to signify the pixel in the image and the weights for bilinear interpolation. Finally, the voxel value v is needed as well. Representing v using 8 bits all four values together can be stored as $x^{11.13b}u^{8.0b}y^{11.13b}v^{8.0b}$, which occupies a cache friendly total of 64 bits. After a table lookup, only a single shift gets x or y into a processor register and a single masking AND to get u or v into a register.

5. Results

In summary, the minimum data precision requirements are as follows:

rendering stage	input	output
sample locations	n/a	11.12b
sample interpolation	12.00b	12.12b
classification	12.00b	0.15b
gradients	12.12b	1.12b
shading	1.12b	1.15b
compositing	1.15b	1.15b

For texture map volume rendering all data items have to be reduced to 8 bits and speculr reflections can not be computed. However, for the restricted case of visualizing one or possibly multiple opaque iso-surfaces, the OpenGL and DirtectX rendering pipelines do provide sufficient precision.

Note, that the output data may be stored either as a single value or in an image-sized buffer. For raycasting, unless a ray-beam is traced (for tile-casting), all output items are computed on a pixel-basis and no buffers are needed. On the other hand, for the image-aligned sheet-buffered splatting algorithm both sample interpolation and compositing occurs on a sheet-basis. However, all other values can be computed on a pixel-basis. For rendering with slice-based texture mapping hardware a similar situation is given. Here, however, all data items have to be reduced to 8 bits and specular reflections can not be computed. Nevertheless, for the restricted case of visualizing one or possibly multiple iso-surfaces the OpenGL and DirtectX rendering pipelines do provide sufficient precision.

Modern PC CPUs are highly optimized to use integers that exactly fit the internal registers. A Pentium 4 can perform two simple 32-bit integer operations per clock cycle. Our analysis showed, that we never needed more than 16-bit precision for the integer part of our computations and also never more than 16 bits of precision for the fractional part. Hence, everything conveniently fits into a 16.16b fixed-point format, which also happens to be the ideal fit considering hardware resources. We measured speedups of up to 4$\times$ compared to using 32-bit floating-point format.

It should be noted that fixed-point arithmetic is not always faster than floating-point arithmetic on today's CPUs. For example, while additions are faster for integer-based fixed-point numbers, multiplications tend to be slower due the fact that integers use the floating-point unit as well, but in addition incur an extra shift operation. At the current date, we do not know if the same holds true for the upcoming GPU and CPU generations. In practice, we use floating-point arithmentic for all color channel operations, such as classification, gradients, shading, and compositing, while we use fixed-point arithmentic for all geometry operations, such as viewing transform, splat positioning, and rasterization. The floating-point format will use 8 bits to store the exponent and one bit for the sign, which leaves 23 bits precision for the narrow dynamic range volume rendering is using. Hence, both formats will suffice for our precision requirements, and we can simply pick the format that yields best performance in the CPU or GPU.

6. Conclusions

We have shown how the physical display hardware limitations of 8 bits per color channel propagate back through the rendering pipeline and how each pipeline stage can be optimized for its numerical precision requirements. The union of all rendering pipeline stage requirements is captured in the 32-bit floating-point format. Double precision does not increase the perceived image quality. Alternatively, a 16.16b format can be used to achieve the same quality images. We also showed that for splatting transformed voxel locations are sufficiently stored in a 64-bit $x^{11.13b}u^{8.0b}y^{11.13b}v^{8.0b}$ format. For splatting and ray casting alike transfer function table entries are sufficiently stored in a 32-bit $R^{0.8b}G^{0.8b}B^{0.8b}\sqrt{\alpha}^{0.8b}$ format and partially composited pixels / fragments are sufficiently stored in a 64-bit $R^{1.15b}G^{1.15b}B^{1.15b}T^{1.15b}$ format.

7. Acknowledgements

This work was partially supported by Hewlett Packard Laboratories, by ONR grant N000140110034, NSF CAREER grant ACI-0093157, and DOE grant MO-068. We would also like to thank Tom Malzbender and Michael Meissner for many fruitful technical discussions.

References

[Bli95] BLINN J. F.: Jim Blinn's corner: Three wrongs make a right. *IEEE Computer Graphics and Applications 15*, 6 (Nov. 1995), 90–93.

[EKE01] ENGEL K., KRAUS M., ERTL T.: High-quality pre-integrated volume rendering using hardware-accelerated pixel shading. In *Eurographics / SIGGRAPH Workshop on Graphics Hardware* (2001), pp. 9–16.

[IL95] IHM I., LEE R.: On enhancing the speed of splatting with indexing. In *Proceedings of Visualization '95* (Oct. 1995), Nielson G. M., Silver D., (Eds.), IEEE, pp. 69–76.

[Kni00] KNITTEL G.: The ultravis system. In *Volume Visualization and Graphics Symposium* (Salt Lake City, Utah, Oct. 2000), IEEE, pp. 71–79.

[Kru91] KRUEGER W.: The application of transport theory to visualization of 3-d scalar data fields. *Computers in Physics 5*, 4 (July 1991), 397–406.

[KV84] KAJIYA J. T., VON HERZEN B. P.: Ray tracing volume densities. In *Computer Graphics, SIGGRAPH* (Minneapolis, Minnesota, July 1984), Christiansen H., (Ed.), vol. 18, pp. 165–174.

[LCCK02] LEVEN J., CORSO J., COHEN J., KUMAR S.: Interactive visualization of unstructured grids using hierarchical textures. In *Volume Visualization and Graphics Symposium* (Boston, MA, Oct. 2002), IEEE, pp. 37–44.

[LCN98] LICHTENBELT B., CRANE R., NAQVI S.: *Introduction to Volume Rendering*. Prentice Hall PTR, 1998.

[Lev88] LEVOY M.: Display of surfaces from volume data. *IEEE Computer Graphics and Applications 8(5)* (May 1988), 29–37.

[LL94] LACROUTE P., LEVOY M.: Fast volume rendering using a shear-warp factorization of the viewing transformation. In *Computer Graphics, SIGGRAPH* (Orlando, FL, July 1994), vol. 28, ACM, pp. 451–458.

[MGS02] MEISSNER M., GUTHE S., STRASSER W.: Interactive lighting models and pre-integration for volume rendering on PC graphics accelerators. In *Proc. Graphics Interface* (May 2002), pp. 209–218.

[MHB*00] MEISSNER M., HUANG J., BARTZ D., MUELLER K., CRAWFIS R.: A practical comparison of popular volume rendering algorithms. In *2000 Symposium on Volume Rendering* (2000), pp. 81–90.

[MJC02] MORA B., JESSEL J.-P., CAUBET R.: A new object-order raycasting system. In *IEEE Visualization 2002* (2002), pp. 203–210.

[MKW*02] MEISSNER M., KANUS U., WETEKAM G., HIRCHE J., EHLERT A., STRASSER W., DOGGETT M., FORTHMANN P., PROKSA R.: Vizardii: A reconfigurable interactive volume rendering system. In *2002 Eurographics/SIGGRAPH Workshop on Graphics Hardware* (2002).

[MMC99] MUELLER K., MÖLLER T., CRAWFIS R.: Splatting without the blur. In *IEEE Visualization '99 Conference Proceedings* (N.Y., Oct. 1999), Ebert D., Gross M.,, Hamann B., (Eds.), IEEE, ACM Press, pp. 363–370.

[MSHC99] MUELLER K., SHAREEF N., HUANG J., CRAWFIS R.: High-quality splatting on rectilinear grids with efficient culling of occluded voxels. *IEEE Transactions on Visualization and Computer Graphics 5*, 2 (Apr.June 1999), 116–134.

[PHK*99] PFISTER H., HARDENBERGH J., KNITTEL J., LAUER H., SEILER L.: The volumepro real-time ray-casting system. In *Computer Graphics, SIGGRAPH* (Los Angeles, CA, Aug. 1999), ACM Siggraph, pp. 251–260.

[PK96] PFISTER H., KAUFMAN A. E.: Cube-4 - a scalable architecture for real-time volume visualization. In *Symposium on Volume Visualization* (San Francisco, CA, Oct. 1996), ACM, pp. 47–54.

[PSL*98] PARKER S., SHIRLEY P., LIVNAT Y., HANSEN C., SLOAN P.-P.: Interactive ray tracing for isosurface rendering. In *IEEE Visualization '98 Conference Proceedings* (Oct. 1998), Ebert D., Hagen H.,, Rushmeier H., (Eds.), IEEE.

[RSEB*00] REZK-SALAMA C., ENGEL K., BAUER M., GREINER G., ERTL T.: Interactive volume rendering on standard pc graphics hardware using multi-textures and multi-stage-rasterization. In *2000 Siggraph/Eurographics Workshop on Graphics Hardware* (2000).

[Wes90] WESTOVER L.: Footprint evaluation for volume rendering. In *Computer Graphics, SIGGRAPH* (Dallas, TX, July 1990), vol. 24(4), ACM, pp. 367–376.

[WMG98] WITTENBRINK C. M., MALZBENDER T., GOSS M. E.: Opacity-weighted color interpolation for volume sampling. In *IEEE Symposium on Volume Visualization* (Chapel Hill, NC, Oct. 1998), IEEE, pp. 135–142.

[WV92] WILHELMS J., VANGELDER A.: Octrees for faster isosurface generation. *ACM Transactions on Graphics 11*, 3 (1992), 201–227.

Graphics Hardware (2004)
T. Akenine-Möller, M. McCool (Editors)

Mio: Fast Multipass Partitioning via Priority-Based Instruction Scheduling

Andrew Riffel[⋆], Aaron E. Lefohn[⋆†], Kiril Vidimce[†], Mark Leone[†], and John D. Owens[⋆]

[⋆]University of California, Davis [†]Pixar Animation Studios

Abstract

Real-time graphics hardware continues to offer improved resources for programmable vertex and fragment shaders. However, shader programmers continue to write shaders that require more resources than are available in the hardware. One way to virtualize the resources necessary to run complex shaders is to partition the shaders into multiple rendering passes. This problem, called the "Multi-Pass Partitioning Problem" (MPP), and a solution for the problem, Recursive Dominator Split (RDS), have been presented by Eric Chan et al. The $O(n^3)$ *RDS algorithm and its heuristic-based* $O(n^2)$ *cousin,* RDS_h*, are robust in that they can efficiently partition shaders for many architectures with varying resources. However, RDS's high runtime cost and inability to handle multiple outputs per pass make it less desirable for real-time use on today's latest graphics hardware. This paper redefines the MPP as a scheduling problem and uses scheduling algorithms that allow incremental resource estimation and pass computation in* $O(n \log n)$ *time. Our scheduling algorithm, Mio, is experimentally compared to RDS and shown to have better run-time scaling and produce comparable partitions for emerging hardware architectures.*

Categories and Subject Descriptors (according to ACM CCS): I.3.2 [Computer Graphics]: Graphics Systems; D.3.4 [Programming Languages]: Processors—Compilers

1. Introduction

Recent advances in architecture and programming interfaces have added substantial *programmability* to the graphics pipeline. These new features allow graphics programmers to write user-specified programs that run on each vertex and each fragment that passes through the graphics pipeline. Based on these *vertex programs* and *fragment programs*, several groups have developed *shading languages* that are used to create real-time programmable shading systems that run on modern graphics hardware.

The ideal interface for these shading languages is one that allows its users to write arbitrary programs for each vertex and each fragment. Unfortunately, the underlying graphics hardware has significant restrictions that make such a task difficult. For example, the fragment and vertex shaders in modern graphics processors have restrictions on the length of programs, on the number of constants and registers that may be accessed in such programs, and the control flow constructs that may be used.

Each new generation of graphics hardware has raised these limits. As an example, Microsoft's DirectX graphics standard specifies the maximum size in instructions of vertex and fragment (pixel) shader programs, and graphics hardware has closely tracked these limits. For instance, the vertex shader 1.0 specification allowed 128 instructions, and the 2.0 and 3.0 specifications increased this limit to 256 and 512 instructions, respectively. Fragment programs have increased in instruction size even more quickly: while version 1.3 pixel shaders could only run 12 instructions, version 2.0 increased this limit to 96 instructions, and version 3.0 supports a minimum of 512 instructions (with looping allowing a dynamic instruction count of up to 65,536).

This rapid increase in possible program sizes, coupled with parallel advances in the capability and flexibility of vertex and fragment instruction sets, has led to corresponding advances in the complexity and quality of programmable shaders. For many users, the limits specified by the latest standards already exceed their needs. However, at least two major classes of users require substantially more resources for their applications of interest, and it is these users whom we target with this work.

The first class of users we target are those who require shaders with more complexity than current hardware can support. Many shaders in use in the fields of photorealistic rendering or film production, for instance, exceed the capabilities of current graphics hardware by at least an order of magnitude. The popular RenderMan shading language [Ups90] is often used to specify these shaders, and RenderMan shaders of tens or even hundreds of thousands of instructions are not uncommon. Implementing these complex RenderMan shaders is not possible in a single vertex or fragment program.

The second class of users we target use graphics hardware to implement general-purpose (often scientific) programs. This "GPGPU" (general-purpose on graphics processing units) community targets the programmable features of the graphics hardware in their applications, using the inherent parallelism of the graphics processor to achieve superior performance to microprocessor-based solutions. Like complex RenderMan shaders, GPGPU programs often have substantially larger programs than can be implemented in a single vertex or fragment program. They also may have more complex outputs; instead of a single color, they may need to output a compound data type.

To implement larger shaders than the hardware allows, programmers have turned to *multipass methods* in which the shader is divided into multiple smaller shaders, each of which respects the hardware's resource constraints. These smaller shaders are then mapped to multiple passes through the graphics pipeline; each pass produces results that are saved for use in future passes.

The key step in this process is the efficient partitioning of the shader into several smaller shaders. Chan et al. call this the "multipass partitioning problem". In their 2002 paper, they propose a solution to this problem with a polynomial-time algorithm entitled "Recursive Dominator Split" (RDS) [CNS*02]. Currently, RDS is the state of the art in solving the multipass partitioning problem. ATI's prototype Ashli compiler [BP03], for example, uses RDS to partition its input RenderMan and GL Shading Language shader programs into multiple passes for use on graphics hardware. The OpenGL Shading Language is an example of a language that requires solutions to the multipass partitioning problem such as RDS: it calls for shading language implementations to "virtualiz[e] resources that are not easy to count", including the "number of machine instructions in the final executable" [KBR03].

Despite its success, RDS has two major deficiencies that have made it less than ideal for use in modern graphics systems.

- First, shader compilation in modern graphics systems is performed dynamically at the time the shader is run. Consequently, graphics vendors require algorithms that run as quickly as possible. Given n instructions, the runtime of RDS scales as $O(n^3)$. Even the heuristic version of RDS, RDS_h, scales as $O(n^2)$. This high runtime cost makes RDS and RDS_h undesirable for implementation in runtime compilers.
- Second, RDS assumes a hardware target that can output at most one value per shader per pass. While the hardware targeted by RDS at the time conformed to this restriction, more modern hardware allows multiple outputs per pass, and the RDS algorithm does not handle this capability. Because allowing more outputs in each pass implies more operations that generate those outputs, supporting multiple render targets per pass increases the effective number of operations that can be scheduled in a single pass.

In this work we present Mio, an instruction scheduling approach to the multipass partitioning problem that addresses these two deficiencies. Our algorithms have better complexity than RDS ($O(n \log n)$ in the average case and $\Omega(n^2)$ in the worst case) and efficiently utilize multiple outputs per pass. In addition, though we do not explore this point in this work, our solutions will continue to be applicable as shader programs use control flow.

The remainder of this paper is organized as follows. In the next section we describe the previous work in this field, and in Section 3 the RDS algorithm in detail. In Section 4, we describe our algorithms and techniques that we have developed to solve the multipass partitioning problem. We explain our experimental setup in Section 5 and our results and discussion in Section 6. Our future work is enumerated in Section 7 and we conclude in Section 8.

2. Previous Work

Shading languages were introduced by Rob Cook in his seminal paper on shade trees [Coo84], which generalized the wide variety of shading and texturing models at the time into an abstract model. He provided a set of basic operations which could be combined into shaders of arbitrary complexity. His work was the basis of today's shading languages, including Hanrahan and Lawson's 1990 paper [HL90], which in turn contributed ideas to the widely-used RenderMan shading language [Ups90].

While these shading languages were not meant for interactive use, several recent languages target real-time shading applications. Olano and Lastra developed a shading language for the PixelFlow graphics system in 1998; three years later, Proudfoot et al. targeted the programmable stages of NVIDIA and ATI hardware with their Real-Time Shading Language (RTSL) [PMTH01]. Many of RTSL's ideas were incorporated into NVIDIA's Cg shading language [MGAK03], also aimed at modern graphics hardware. Other notable shading languages for today's graphics processors include SMASH [McC01] and the OpenGL Shading Language [KBR03].

Peercy et al. implemented a shading compiler on top of OpenGL hardware [POAU00]; they characterized OpenGL as a general SIMD computer and mapped RenderMan shaders to this SIMD framework. To map complex shaders to the multiple OpenGL passes required to implement these shaders, they used compile-time dynamic tree-matching techniques. SGI offers a product based on this technology (featuring both compile-time and runtime compiler phases) called OpenGL Shader [SGI04].

3. Recursive Dominator Split (RDS)

The work of Chan et al. presented runtime techniques to map complex fragment shaders to multiple passes of modern programmable graphics hardware [CNS*02]. Their Recursive Dominator Split (RDS) algorithm generates a solution to the multipass partitioning problem. RDS is successful at finding near-optimal partitions for a variety of shaders on a diversity of cost models. Their system, implemented on top of the Stanford Real-Time Programmable Shading System, used a cost function incorporating the number of passes, the number of texture lookups, and the number of instructions, and then optimized the resulting multipass mapping to minimize this cost function. They described two particular algorithms: RDS, which uses search to make its save/recompute decisions, and RDS_h, which uses a heuristic instead. They conclude that RDS_h runs faster than RDS at the expense of partition quality; also, they find that optimal partitions generated through exhaustive search are only 0–15% better than those generated by RDS (which, in turn, are 10.5% better than those generated by RDS_h).

We first describe the assumptions made by RDS that motivate its design, then describe the algorithm itself.

3.1. Assumptions

Minimization Criteria RDS's primary goal in dividing a shader into multiple passes is to minimize the number of passes. This decision is motivated by the overhead incurred on each pass, which includes the cost of setting up and running the input through all stages of the graphics pipeline as well as the runtime and memory cost of texture storage of intermediate results. Consequently, Chan et al. indicate that "RDS was designed under the assumption that per-pass overhead dominates the overall cost."

Save vs. Recompute One core decision that RDS considers is how to handle multiply-referenced nodes in the shade tree. If a node is multiply referenced, RDS must make a decision whether to compute and *save* the value at that node for use by its multiple references, or instead whether to *recompute* the value of that node at each of its multiple references. Saving reduces the number of overall instructions; recomputing reduces the number of passes at the cost of an increase in the number of instructions.

Structure of Shader Input RDS assumes its input shader is expressible as a directed acyclic graph (DAG).

Outputs per Pass RDS assumes a single output per pass, and that output can only come from a single node (though the shader hardware targeted by RDS actually outputs a 4-component vector per pass, RDS would not be able to combine a 3-component result from one operation with a 1-component result from a different operation in a single pass's output).

3.2. The RDS Algorithm

RDS begins by identifying multiply-referenced nodes in its input DAG and deciding whether the values generated at those nodes should be produced in a separate pass and saved or instead recomputed. It then starts at the leaves of the DAG and traverses the DAG in postorder, greedily merging nodes into as few passes as possible.

Chan et al. present two algorithms to make the decision to save or recompute. The first, RDS_h, uses a heuristic to make this decision: the value at the multiply-referenced node is recomputed only if the resource consumption in the recomputation uses less than half the maximum resources allowed (operations, registers, etc.) in a single pass. The second, RDS, combines RDS_h's save-recompute heuristic with a limited search over evaluations of partitions created by both saving and recomputing multiply-referenced nodes.

3.2.1. Cost of RDS

The runtime cost of both RDS and RDS_h depends on the cost to check if a subregion can be mapped to a single pass; for n nodes in the input DAG, Chan et al. define this cost as $g(n)$ and assume this cost is $O(n)$. They then demonstrate that RDS scales as $O(n^2) \cdot g(n) = O(n^3)$ and that RDS_h scales as $O(n) \cdot g(n) = O(n^2)$.

3.2.2. Limitations in RDS

The RDS algorithm has three major limitations. First, its runtime cost of $O(n^3)$ (or $O(n^2)$ for RDS_h) makes it undesirable for use in runtime compilers. Next, because RDS considers only a single connected region of its input DAG at any time, it cannot handle more than one output per pass (and is not easily extended to do so within its $O(n^3)$ runtime cost; in concurrent work, Foley et al. demonstrate a $O(n^4)$ algorithm that can spill multiple outputs per pass [FHH04]). Finally, because RDS assumes its input is structured as a single DAG, it cannot support flow control operations such as conditionals or loops. Recent hardware supports both multiple outputs per pass as well as control flow within a shader, so while these two limitations were not restrictions in 2002 hardware, they will be in the hardware of today and tomorrow.

4. Mio: An Instruction Scheduling Algorithm

In this paper we present an alternate algorithm to solve the multipass partitioning problem. Our solution is motivated by two goals. First, the limitations of RDS described above do not allow RDS-produced shaders to take advantage of important features of modern graphics hardware. Second, the runtime cost of RDS makes it unsuitable for use in the dynamic compilation environments used by the drivers of modern GPUs, as well as problematic for partitioning high-quality rendering or GPGPU shaders with thousands of instructions. To address these two deficiencies, we present our solution, called "Mio" (pronounced "mee-oh"). The name Mio is derived from the word "meiosis," a process of cell division that produces child cells with half the number of chromosomes as their parents. In the same way, Mio divides large programs into smaller partitions.

We begin by contrasting the assumptions made in the design of RDS (Section 3.1) to the assumptions we have made in the design of Mio. We next describe the technique of "list scheduling" that is the core of Mio's scheduling algorithm, then describe Mio and its operation.

4.1. Assumptions

Minimization Criteria Each recent generation of graphics hardware has allowed shader programs with greater resource limits than the previous generation. Specifically, the newest hardware can run more operations, access more textures, and use more constants and registers than the generation before.

These increasing resource limits cause a shift in our minimization goal. While RDS attempts to minimize the number of *passes* in its solution, often at the expense of additional operations to recompute an intermediate result, Mio instead attempts to minimize the number of *operations*. RDS's decision was valid given the characteristics of the hardware at the time: the hardware targeted by RDS, the ATI Radeon 8500, allowed only 16 operations on each pass.

With a larger number of operations allowed per pass, however, the overhead per pass becomes less significant. As the number of operations allowed per pass continues to increase, optimizing shader runtime no longer requires minimizing the number of passes but instead becomes minimizing the number of operations. We show in Section 6.1 that the performance of shaders on modern graphics hardware is no longer limited by the number of passes but instead by the number of operations.

Save vs. Recompute For a node with a value that is multiply-referenced, RDS decides whether to save or recompute that node's value for each reference. Because we have chosen to optimize for the number of operations, we never recompute the value at a multiply-referenced node.

Structure of Shader Input Like RDS, our current implementation of Mio requires that its input shaders be expressible as DAGs. List scheduling can potentially handle more complex input, however, which we briefly discuss in Section 7.

Outputs per Pass In contrast to RDS, Mio accounts for both scalar (1-component) and vector (4-component) datatypes and coalesces multiple operation results into a single output when appropriate. For instance, for graphics hardware with 4 4-vector outputs, Mio supports up to 16 scalar outputs. We make two explicit assumptions in determining the outputs per pass:

1. We prefer to fill all 4 entries in an output vector when possible.
2. We also prefer to use more available outputs per pass when possible.

Together, these two assumptions imply that we use the maximum resources allowed by the hardware to communicate results between passes. Since the limited amount of these resources is often the limiting factor in creating efficient partitions, our assumptions allow us to create schedules with lower operation and pass counts than if we do not use all these resources. However, doing so may incur the cost of more memory bandwidth when compared to less aggressive inter-pass data communication. While this additional bandwidth cost may be a possible bottleneck, we note that the number of possible outputs per pass is growing much more slowly than the number of operations per pass, and thus the relative spill cost is decreasing with time.

These assumptions also imply that the graphics hardware can use its memory bandwidth equally efficiently with scalar, 1-output-per-pass outputs and vector, n-outputs-per-pass outputs. Though some hardware has a measured peak bandwidth that declines as the number of render targets increases, the newest NVIDIA hardware appears to maintain a constant bandwidth independent of the number of render targets [Buc04].

4.2. List Scheduling

With this work we redefine the MPP as a Precedence Constrained Job-Shop Scheduling Problem (PCJSSP), a well-known problem in the compiler community. Among the problems analogous to the PCJSSP are instruction scheduling, temporal partitioning, and golf tee time assignment. The input to the PCJSSP is a list of jobs, the resources they require (including time), and the dependencies between them. The goal of the problem is to schedule each job to a shop (resource), respecting resource limitations while minimizing the overall time to complete all the jobs. The PCJSSP is a NP-complete problem, but the technique of *list scheduling* provides good and often near-optimal solutions to the PCJSSP in polynomial time.

List scheduling is a greedy algorithm that tries to schedule jobs based on priority as those jobs become "ready." A job is ready when all of its predecessors are scheduled and the results created by those predecessors are available. The list scheduling algorithm begins with the first time slot, scheduling jobs into that time slot based on job priority, until the resources available during that time slot are exhausted or dependencies prevent the scheduling of any jobs. It then moves to the next time slot(s) and continues to schedule jobs until all jobs are scheduled. At all times it maintains a ready list of jobs. Initially this list consists of all jobs with no dependencies, but as jobs are scheduled, new jobs whose dependencies are now satisfied are added to the ready list.

The effectiveness of list scheduling hinges upon how priorities are assigned to jobs. A bad priority scheme can cause the resulting schedule to be far from optimal. One popular and effective scheme is critical-path priority, which gives preference to jobs that have the most jobs (or time) between them and the final job or jobs. Another scheme with good performance is most-successors priority, which prefers to schedule jobs with the most dependencies as early as possible. Jain and Meeran provide a comprehensive review of job-shop scheduling techniques [JM98]. A representative example of an early analysis of list scheduling is from Landskov et al. [LDSM80]; Cooper et al. evaluate the conditions under which list scheduling performs well and poorly [CSS98]. Inagaki et al. have developed the state of the art in runtime list scheduling techniques that integrate register minimization into scheduling and feature efficient backtracking [IKN03].

We define the MPP as a job-shop problem in the following way. Jobs are defined as shader operations; resources include shader functional units, registers (varying, constant, texture, and output), and slots in shader instruction memory. The MPP differs from traditional instruction scheduling in two important ways. First, traditional job-shop instruction scheduling considers functional units as the primary resource and does not consider register allocation, whereas in the MPP we must consider not only functional units as resources but also many other resources (registers, outputs, textures, etc.). Second, to solve the MPP we must generate outputs that map to passes in graphics hardware. In the context of traditional instruction scheduling, we must produce a schedule that can be segmented into multiple passes; this requires that no more than k registers can be live at the breakpoint between passes, where k is the number of allowed outputs for each pass of the shader. These k results can then be saved into textures and used in future passes.

4.3. The Mio Algorithm

Mio is an instruction-scheduling approach to the MPP. Mio's goal is to schedule operations into the minimum amount of total time while respecting the limitations of the underlying hardware. Traditional list scheduling methods for modern superscalar microprocessors attempt to maximize the amount of instruction-level parallelism in their programs. To do so, they attempt to maximize the number of operations in their ready list, thus allowing a maximal opportunity for parallel instruction issue. This strategy motivates a "wide" traversal of their input DAGs, working on many subtrees at the same time. Modern microprocessors feature large register files and register-renaming hardware to store the large number of temporary results that result from this traversal.

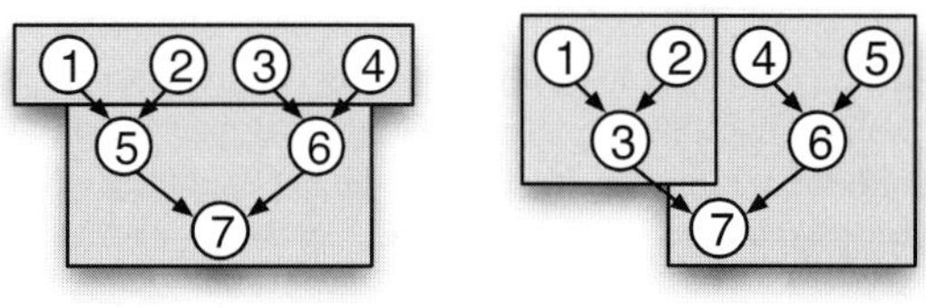

(a) Traditional. (b) Mio.

Figure 1: *Traditional list scheduling priority schemes attempt to maximize the parallelism in their instruction stream. Mio, on the other hand, minimizes register usage. Here we consider a 7-operation DAG that must be partitioned into two passes. Note that the traditional method can issue all 4 operations in its first pass in parallel but must pass 4 temporaries between passes. Mio, on the other hand, chooses a different set of operation priorities to only require 1 temporary between its passes.*

Modern GPUs, on the other hand, have a severe limitation in their ability to store intermediate results across multiple passes. They are constrained by the number of rendering targets and the width of storage to each rendering target—ATI's X800 and NVIDIA's GeForce 6800, for instance, support fragment programs of hundreds of unique operations but can only store 4 4-vectors at the end of each pass. The "wide" traversal typical of today's common list scheduling implementations is therefore poorly suited to graphics hardware because of the GPU's limited ability to store intermediates.

Instead, a list scheduling algorithm for GPUs must perform a "deep" traversal of its input DAG (Figure 1). The deep traversal attempts to wholly traverse a given subtree before another subtree is begun. This traversal produces an instruction schedule with less register usage at the cost of instruction parallelism. We now describe the Mio algorithm and its priority scheme.

- We begin with two passes through the input to generate the traversal order $\mathcal{T}$ for the DAG. The first pass assigns a Sethi-Ullman number to each node of the DAG, which indicates the number of registers required to execute the subtree rooted at that node. For tree-structured inputs, Sethi and Ullman have demonstrated that scheduling higher-numbered nodes first minimizes overall register usage [SU70]. (The corresponding problem for DAGs is NP-complete; Aho gives heuristics for computing good DAG orderings [AJU77], but we have found simply using the Sethi-Ullman numbering on input DAGs gives acceptable results in our tests.) This pass is $O(n)$ with the number of input nodes.
- On our second pass, we visit each node using a postorder

depth-first traversal through the input DAG, preferentially choosing the node with the highest Sethi-Ullman number. We break ties in an arbitrary manner (in our implementation, we use a pointer comparison). This pass is also $O(n)$ with the number of input nodes, and at its end, we have a traversal order $\mathcal{T}$ for all operations in the input DAG.

After the two prepasses, we can now schedule operations.

- We first insert all operations with no predecessors into the ready list.
- Next, all operations in the ready list are considered for scheduling. The ready list is implemented as a priority queue and we choose the operation in the ready list with the highest priority. The priority is simple: nodes (representing operations) earlier in the traversal order $\mathcal{T}$ are preferred over nodes later in $\mathcal{T}$.
- We then attempt to schedule the highest priority node from the ready list. An incremental resource estimator determines if the node can be added to the current pass without violating hardware constraints (such as the number of temporary registers, textures, etc.). If the resource estimator rejects a node because of the number of operations or temporary registers (hard limits that affect all operations), we empty the ready list and end the pass. If that limitation is instead an input constraint that may vary depending on which operations are scheduled (textures, varyings, or uniforms), we remove the operation from the ready list but continue to attempt to schedule other operations with the hope that they will not violate the constraint. (One possible optimization that we do not implement is to increase the priority of operations that do not increase a critical resource when that resource is nearly exhausted.) We do allow the scheduling of an operation that temporarily overutilizes the number of outputs per pass with the hope that future operations will return the schedule to compliance.
- After we successfully schedule an operation, we add its successor operations that no longer have any unscheduled predecessors to the ready list and begin scheduling the next highest priority operation.
- If the ready list is empty, we have completed a pass. We then check if the number of outputs exceeds the hardware constraint. If so, we require a rollback to a point where the constraints of the hardware are met. To facilitate this rollback when necessary, after every operation is scheduled, we check to see if we violate the output constraint. If we do, we add the current operation to a rollback stack. We clear the stack if we later add a node that puts the scheduled pass back into compliance with the output constraint. If we reach the end of the pass and do not meet the constraints, we must unschedule each node on the stack to roll back to a schedule that does meet the constraints.

One of the advantages of the list scheduling formulation is that any priority scheme can be used, not just the Sethi-Ullman numbering we use here. More sophisticated and accurate priorities could easily be integrated into the framework we have developed.

4.3.1. Analysis of Mio Runtime

The major costs in using Mio to partition an input shader into multiple passes are the cost of the two prepasses, the cost of scheduling of each operation, the cost of the maintenance of the ready list, and the cost of rollback.

Both prepasses are $O(n)$ since each node must be visited only once. If we assume the cost of maintaining the ready list is $r(n)$, with no rollback all operations can be scheduled in $O(n)\cdot r(n)$. Most implementations of a ready list take $O(\log n)$ time; we implement the ready list as a $O(\log n)$ priority queue (specifically a heap). Though Gabow and Tarjan have demonstrated that the overhead of a priority queue can be reduced to $O(1)$ [GT83], the constants associated with their method are large enough that only extremely large shaders would run more efficiently than with $O(\log n)$ methods. Overall, then, with no rollback the expected runtime of Mio is $O(n \log n)$.

Rollback adds an additional $O(n)$ cost to the cost of scheduling all operations (assuming a rollback at each operation), making the worst-case overall cost $\Omega(n^2)$. (Note that it is only necessary to calculate the priority queue once per node rather than once per visit to the node, so the cost is not $\Omega(n^2 \log n)$.) It is important to note, however, that we do not continue to schedule in the current pass after rolling back, so the number of rollbacks can never be more than the number of passes. As the number of outputs per pass increases, rollbacks become less frequent and the cost of rollback consequently decreases. Section 6.2 shows our results for compiler runtime.

When resource constraints are violated, rollback is not the only option. We could instead change the priority scheme and attempt to repair the schedule, but this would likely impose a higher runtime cost in exchange for a potentially better schedule (and a greater algorithmic complexity). We could also consider priority schemes that analyze resource usage for subsets of the input DAG instead of only one operation at a time, but again, these schemes would impose additional runtime costs.

5. Experimental Setup

We integrated Mio into ATI's prototype Ashli compiler [BP03]. Ashli implements RDS_h and all results in this paper are compared against Ashli's RDS_h implementation.

To evaluate the effectiveness of Mio, we measured its performance on a variety of RenderMan shader programs. Both RDS_h and Mio share the Ashli front end, which parses the shader into the DAG used by the RDS_h and Mio back ends. We have attempted to select shaders that represent a variety of fragment program types, including ones with ample and

Shader	WOOD	MATTE	POTATO
Uniform	10	2	13
Varying	4	4	6
Textures	1	0	1
Operations	67	9	148
Shader	**BAT**	**UBER**	**MONDO**
Uniform	7	36	128
Varying	0	4	4
Textures	0	3	41
Operations	352	88	559

Table 1: *Shader Characteristics. Our shaders of interest are listed with their relevant resource demands and operation counts.*

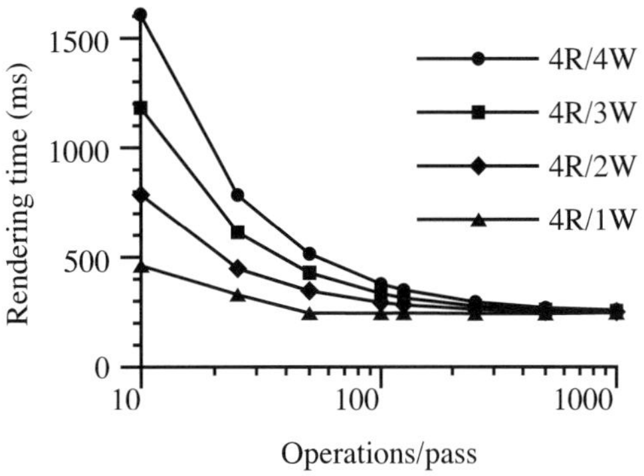

Figure 2: *Runtime of a 5000-instruction shader for 4 reads/n writes (n = 1, 2, 3, and 4) as we vary the number of operations per pass.*

little texture usage and ones with many and few multiply-referenced nodes.

We use three surface shaders in our tests. The WOOD shader simulates wood grain; it is the same core shader used by Chan et al. [CNS*02]. MATTE is a simple matte surface shader. POTATO is a more complicated surface shader that approximates a potato skin. We also use three lights. BAT outlines the silhouette of a bat within a spotlight. The UBERLIGHT (UBER) shader implements a complex lighting model representative of those used in computer-generated film production. The shader we use here is a simplification of Gritz's implementation of Barzel's lighting model [Bar97]. We augment UBER with 10 fake shadows, 15 shadow maps, and 25 slides to create MONDOLIGHT (MONDO), whose texture use is large enough to require multiple passes on even Pixel Shader 3.0 hardware. In each of our tests we combine one surface shader with *N* lights (designated as SURFACE-LIGHTN). Relevant shader characteristics are summarized in Table 1.

We report results on two hardware configurations, Mio48 and Mio64, that target ATI Radeon 9800 Pro limits. Mio48 allows 48 ALU and 32 texture operations per pass, while Mio64 allows 64 and 32. It is necessary to use these smaller limits for the results we present in the next section to create a more difficult partitioning problem due to smaller partitions, to allow runtime measurements on the ATI card, and to permit a comparison with Ashli's implementation of RDS_h, which does not yet handle larger shaders. Because of an inefficiency in the Ashli code generator (all results must be explicitly copied to output registers), Mio48 represents a lower bound for Mio performance (16 operations are reserved for copying results to pass outputs per pass, so only 48 operations are usable) and an upper bound, Mio64 (in which all 64 operations can be used for useful work). We expect that a more efficient code generator would put optimized results closer to Mio64. When compared to Mio48 and Mio64, all RDS_h results allow 64 ALU and 32 texture operations per pass.

Compile-time and runtime results were generated on a 2.8 GHz Pentium IV system running Microsoft Windows XP with 1 GB of RAM. Compile-time measurements record only the partitioning part of the compiler. Runtime results were run on a pre-release NVIDIA GeForce 6800 (NV40) graphics card with 256 MB of memory and driver version 61.32 and a ATI Radeon 9800 Pro with 128 MB of RAM and the Catalyst 4.5 driver. All code is written in C++ and compiled using the Microsoft .NET 2003 compiler.

6. Results and Discussion

To demonstrate the utility of Mio, we must answer four questions. First, we show that the assumptions we have made about the limitations of multipass performance are valid in Section 6.1. Next, in Section 6.2, we compare the compiler runtime of Mio to that of RDS_h. We then analyze Mio's generated code statically in Section 6.3 and the runtime of that code in Section 6.4.

6.1. Limitations of Multipass Performance

Splitting a shader into multiple passes requires storing intermediate results at the end of each pass. These intermediate results can then be used in future passes. Because of the cost of each pass, RDS attempts to minimize the number of passes. In Section 4.1, we argue that the greater amount of resources available in modern GPUs motivates minimizing the number of operations instead.

To test this hypothesis, we consider the relationship between shader size and shader runtime. We expect that shaders with a small operation count will be limited by pass overhead, while shaders with a large number of operations will instead be limited by the operations. We measure this relationship by constructing an artificial shader benchmark

Figure 3: *Compiler runtime of* WOODUBER*N (N = 1–10) for* RDS_h *and Mio. The upper and lower bounds for Mio (Mio64 and Mio48) are nearly identical.*

and varying the number of operations allowed per pass. Figure 2 shows this relationship.

In this experiment, we render a 512×512 quad with a 5000-line fragment shader. The shader uses alternating DOT and ADD instructions and three total registers. On each pass we write to n 4-vectors of 32-bit floats (4 tests with n = 1, 2, 3, and 4). We switch between pbuffer surfaces between passes with no context switch and average the results over 20 consecutive tests.

From this experiment we can see that for small numbers of operations per pass, the performance is highly dependent on the number of passes. This agrees with RDS's assumption that minimizing the number of passes is paramount for the small operations per pass characteristic of the hardware targeted by RDS. However, as the number of operations per pass becomes larger, the pass overhead is no longer a factor. The breakpoint for this changeover appears to be roughly 100–200 operations, after which Mio's assumption of minimizing operations rather than passes becomes more relevant.

6.2. Compiler Runtime

In Section 4.3.1 we asserted that the superior theoretical compile-time performance of Mio would result in faster compile times for shaders on Mio than on RDS_h. Figure 3 shows the comparison for a set of WOOD shaders lit with N UBER lights for passes limited to 64 operations. We note that even the lower bound to Mio performance, Mio48, is superior to RDS_h's runtime, and that RDS_h's runtime grows more quickly as the problem size increases. For POTATOBAT5, the compile time for Mio48, Mio64, and RDS_h, respectively, was 0.35, 0.34, and 0.79 seconds; for MATTEMONDO2, 3.29, 3.22, and 3.58 seconds.

6.3. Static Analysis of Generated Shaders

We next analyze the generated code from Mio and RDS_h. Comparing the number of passes and the total number of generated operations allows us to draw conclusions about the quality of the code generated by each algorithm.

Shader	ALU Ops	Tex Ops	Total Ops	Passes
WOODUBER1				
RDS_h	231	12	243	7
Mio48	243	18	261	5
Mio64	219	15	234	4
WOODUBER10				
RDS_h	1836	106	1942	48
Mio48	1870	200	2070	48
Mio64	1632	166	1798	40
MATTEMONDO2				
RDS_h	2853	197	3050	56
Mio48	2738	395	3133	88
Mio64	2347	368	2715	79
POTATOBAT5				
RDS_h	1106	55	1161	25
Mio48	1146	144	1290	37
Mio64	996	130	1126	32

Table 2: *Static results of generated code.*

The results for the WOODUBERN, MATTEMONDO2, and POTATOBAT5 shaders are shown in Table 2. Static Mio operation counts are comparable to RDS_h operation counts, with Mio generally having more texture operations (due to more spills) and fewer arithmetic operations (due to no recomputation). Shaders compiled by Mio also generally have a comparable number of passes to RDS_h. The most fragile part of our implementation is in the rollback decisions, and we believe rollback represents the largest potential area for improvement in these results.

6.4. Runtime Analysis of Generated Shaders

The ultimate test for a compiler is the execution time of its generated code. We use the internal frame counter of the Ashli viewer to measure frame rate results. Unfortunately, the Ashli viewer can only display shaders of a limited size, but we expect the static code results from Section 6.3 to be indicative of runtime performance.

With small shaders with few partitions, we found equal performance between shaders partitioned with RDS_h and with Mio48. For example, running MATTEUBER1 (4 passes using 3 intermediate texture storage buffers with RDS_h, 3 passes using 4 buffers with Mio48), we measured 60 frames per second on the NVIDIA GeForce 6800 and 40 frames per second on the ATI Radeon 9800 Pro. However, the Mio48 partition's greater use of intermediate texture storage, due to spilling to multiple render targets, caused a substantial hit in performance when its use of buffers was considerably greater than RDS_h's partition. For example, adding a second

UBERLIGHT to MATTEUBER1 resulted in Mio's partition using 6 passes and 10 buffers while RDS_h's had 7 passes and 5 buffers. The RDS_h partition ran more than twice as fast on the GeForce 6800 (30 frames per second against 12 for Mio48). We believe that runtime performance is correlated with the number of intermediate buffers but have not yet fully investigated why this is the case. We believe a significant reason for the slowdown we see with a large number of buffers is a buffer's large memory footprint: a 512 $\times$ 512, 4-surface pbuffer takes 17 MB of texture memory, and for Mio partitions that use many buffers, we expect the lower performance is caused by texture memory thrashing. We note that allowing more operations per pass will reduce the relative cost of the spill and so we expect Mio's runtime results will improve as we schedule to hardware with more resources; we describe other possible optimizations in the next section.

7. Future Work

Mio marks only one point on what we hope will be a continuum of partitioning algorithms with tradeoffs between runtime and partition quality. On one hand, faster runtime algorithms ($O(n)$) are desirable for graphics systems that perform all compilation at runtime. On the other hand, more expensive algorithms that produce near-optimal partitions are equally useful in environments that permit a significant amount of compile-time work. We hope that the development of this family of algorithms will allow us to answer the question of whether it is better to continue to grow resource limits with each generation of graphics hardware or whether we should require the graphics system to virtualize the hardware and hide the need for multipass or other methods.

We have begun to explore $O(n)$ partitioning algorithms by removing the ready list and simply choosing the highest node in the Sethi-Ullman order; our initial results are promising, but a much more thorough analysis is required. For more complex algorithms, we will incorporate other information that is not currently used in generating our partitions. First, we need a more sophisticated cost function for spilling results at the end of a pass. Unlike RDS, Mio assumes all operations have the same cost, and that we should use all spill resources if possible, but we believe that judiciously reducing the amount of spilling at appropriate times will result in better runtime results. (For example, slightly increasing the number of passes and operations while significantly reducing the number of spilled results at the end of passes will almost certainly be an overall performance win.)

We would also like to investigate alternate priority schemes within the list scheduling framework. How can we augment or replace Sethi-Ullman numbering to achieve better compile-time performance and code quality? Another interesting problem is the need to consider a higher level of scheduling: we should explicitly schedule *passes* in a multipass shader in order to maximize cache performance and minimize the live memory footprint at any time. In addition, revisiting our recompute vs. save decision would be worthwhile: are there ever circumstances in practice for which recomputing would produce superior results?

We are also interested in performing a more detailed study on the impact of multiple render targets. Informally we note a substantial increase in Mio's partition quality and an equally substantial decrease in runtime as the number of render targets increases. (Simply allowing multiple render targets consequently makes the MPP a much simpler problem in practice.) For the small number of render targets we investigate here, this makes sense, but how does this trend continue as the number of render targets continues to increase?

One noteworthy feature of list scheduling is its ability to handle inputs with flow control [BR91]. The newest shader instruction sets feature loops, branches, and procedure calls, all features that we do not consider in this work. We hope to adapt our list scheduling algorithm to efficiently handle shaders that use these control flow constructs.

8. Conclusion

The major contributions of this work are the characterization of the multi-pass partitioning problem as an instruction scheduling framework and the development of a priority scheme for list scheduling that enables the generated schedules to best match the abilities of the graphics hardware.

We show that Mio, the partitioning algorithm we develop in this paper, allows both better compile-time performance of the shader compiler as well as comparable partitions to the previous best method, RDS, all while supporting multiple rendering targets. The list scheduling techniques at the core of Mio will also be well-suited to more complex shaders in the future, shaders that require more resources from the graphics hardware and that use more complex control-flow structures. We hope this work spurs the development of high-performance, robust virtualizing shader compilers that will fully hide the multi-pass partitioning problem from shader developers and allow even the largest and most complex shaders to be efficiently mapped to graphics hardware.

Acknowledgements

Arcot Preetham and Mark Segal from ATI provided access to the Ashli codebase that we used as the substrate for our implementation. Arcot and Craig Kolb from NVIDIA both provided valuable technical assistance. We would also like to thank Brian Smits, Alex Mohr, Fabio Pellacini and the rendering research group at Pixar for their support and contributions to this work.

Aaron Lefohn is supported by a National Science Foundation Graduate Fellowship. Funding for this project is through UC Davis startup funds and through a generous grant from ChevronTexaco.

References

[AJU77] AHO A. V., JOHNSON S. C., ULLMAN J. D.: Code generation for expressions with common subexpressions. *Journal of the ACM 24*, 1 (1977), 146–160.

[Bar97] BARZEL R.: Lighting controls for computer cinematography. *Journal of Graphics Tools 2*, 1 (1997), 1–20. `http://www.renderman.org/RMR/Shaders/BMRTShaders/uberlight.sl`.

[BP03] BLEIWEISS A., PREETHAM A.: Ashli—Advanced shading language interface. *ACM Siggraph Course Notes* (2003). `http://www.ati.com/developer/SIGGRAPH03/AshliNotes.pdf`.

[BR91] BERNSTEIN D., RODEH M.: Global instruction scheduling for superscalar machines. In *Proceedings of the ACM SIGPLAN Conference on Programming Language Design and Implementation* (1991), pp. 241–255.

[Buc04] BUCK I.: GPGPU: General-purpose computation on graphics hardware—GPU computation strategies & tricks. *ACM Siggraph Course Notes* (Aug. 2004).

[CNS*02] CHAN E., NG R., SEN P., PROUDFOOT K., HANRAHAN P.: Efficient partitioning of fragment shaders for multipass rendering on programmable graphics hardware. In *Graphics Hardware 2002* (Sept. 2002), pp. 69–78.

[Coo84] COOK R. L.: Shade trees. In *Computer Graphics (Proceedings of SIGGRAPH 84)* (July 1984), vol. 18, pp. 223–231.

[CSS98] COOPER K. D., SCHIELKE P. J., SUBRAMANIAN D.: *An Experimental Evaluation of List Scheduling*. Tech. Rep. 98-326, Department of Computer Science, Rice University, Sept. 1998.

[FHH04] FOLEY T., HOUSTON M., HANRAHAN P.: Efficient partitioning of fragment shaders for multiple-output hardware. In *Graphics Hardware 2004* (Aug. 2004).

[GT83] GABOW H. N., TARJAN R. E.: A linear-time algorithm for a special case of disjoint set union. In *Proceedings of the Fifteenth Annual ACM Symposium on Theory of Computing* (1983), pp. 246–251.

[HL90] HANRAHAN P., LAWSON J.: A language for shading and lighting calculations. In *Computer Graphics (Proceedings of SIGGRAPH 90)* (Aug. 1990), vol. 24, pp. 289–298.

[IKN03] INAGAKI T., KOMATSU H., NAKATANI T.: Integrated prepass scheduling for a Java just-in-time compiler on the IA-64 architecture. In *Proceedings of the International Symposium on Code Generation and Optimization* (2003), pp. 159–168.

[JM98] JAIN A. S., MEERAN S.: *A State-of-the-Art Review of Job-Shop Scheduling Techniques*. Tech. rep., Department of Applied Physics, Electronics and Mechanical Engineering, University of Dundee, Scotland, 1998.

[KBR03] KESSENICH J., BALDWIN D., ROST R.: The OpenGL Shading Language version 1.051. `http://www.opengl.org/documentation/oglsl.html`, Feb. 2003.

[LDSM80] LANDSKOV D., DAVIDSON S., SHRIVER B., MALLETT P. W.: Local microcode compaction techniques. *ACM Computing Surveys 12*, 3 (1980), 261–294.

[McC01] MCCOOL M.: *SMASH: A Next-Generation API for Programmable Graphics Accelerators*. Tech. Rep. CS-2000-14, Department of Computer Science, University of Waterloo, 20 April 2001. API Version 0.2.

[MGAK03] MARK W. R., GLANVILLE R. S., AKELEY K., KILGARD M. J.: Cg: A system for programming graphics hardware in a C-like language. *ACM Transactions on Graphics (Proceedings of ACM SIGGRAPH 2003) 22*, 3 (July 2003), 896–907.

[PMTH01] PROUDFOOT K., MARK W. R., TZVETKOV S., HANRAHAN P.: A real-time procedural shading system for programmable graphics hardware. In *Proceedings of ACM SIGGRAPH 2001* (Aug. 2001), Computer Graphics Proceedings, Annual Conference Series, pp. 159–170.

[POAU00] PEERCY M. S., OLANO M., AIREY J., UNGAR P. J.: Interactive multi-pass programmable shading. In *Proceedings of ACM SIGGRAPH 2000* (July 2000), Computer Graphics Proceedings, Annual Conference Series, pp. 425–432.

[SGI04] SGI: OpenGL Shader, 2004. `http://www.sgi.com/software/shader/`.

[SU70] SETHI R., ULLMAN J. D.: The generation of optimal code for arithmetic expressions. *Journal of the ACM 17*, 4 (Oct. 1970), 715–728.

[Ups90] UPSTILL S.: *The Renderman Companion*. Addison-Wesley, 1990.

Graphics Hardware (2004)
T. Akenine-Möller, M. McCool (Editors)

Efficient Partitioning of Fragment Shaders for Multiple-Output Hardware

Tim Foley, Mike Houston and Pat Hanrahan †

Stanford University

Abstract

Partitioning fragment shaders into multiple rendering passes is an effective technique for virtualizing shading resource limits in graphics hardware. The Recursive Dominator Split (RDS) algorithm is a polynomial-time algorithm for partitioning fragment shaders for real-time rendering that has been shown to generate efficient partitions. RDS does not, however, work for shaders with multiple outputs, and does not optimize for hardware with support for multiple render targets.

We present Merging Recursive Dominator Split (MRDS), an extension of the RDS algorithm to shaders with arbitrary numbers of outputs which can efficiently utilize hardware support for multiple render targets, as well as a new cost metric for evaluating the quality of multipass partitions on modern consumer graphics hardware. We demonstrate that partitions generated by our algorithm execute more efficiently than those generated by RDS alone, and that our cost model is effective in predicting the relative performance of multipass partitions.

Categories and Subject Descriptors (according to ACM CCS): I.3.1 [Computer Graphics]: Graphics processors G.2.2 [Mathematics of Computing]: Graph AlgorithmsTrees

1. Introduction

Real-time shading languages for graphics hardware simplify the task of writing shader code that is portable across a range of hardware and graphics APIs. However, most current high-level shading language compilers do not virtualize platform-specific resource limits such as number of instructions, input textures, or render targets. We can virtualize these hardware constraints for fragment shaders with *multipass partitioning* by dividing a large shader into multiple rendering passes.

The goal of multipass partitioning is to generate a set of passes that are equivalent to the input shader such that the cost of executing all the passes on the target hardware is minimized. The size of the search space is exponential in the number of operations, so good heuristic algorithms are necessary to find efficient partitions in reasonable time. The Recursive Dominator Split (RDS) algorithm is a polynomial-time heuristic algorithm for multipass partitioning that generates partitions that execute efficiently for a wide variety of shaders [CNS*02]. However, RDS is limited to operating on shaders with a single output color.

Recently, it has been shown that graphics hardware can also be used to run a large number of non-shading algorithms including ray tracing [PBMH02], fluid dynamics [HBSL03], and stream processing based applications [BFH*04]. These applications frequently require multiple outputs per pass, and hardware support for multiple render targets (MRT) makes it possible to more efficiently execute such shaders.

This paper presents Merging Recursive Dominator Split (MRDS), an extension of the RDS algorithm to support multiple-output shaders and graphics hardware with multiple render targets. The primary contributions of this paper are:

- The extension of the RDS algorithm to support shaders with multiple outputs. Our algorithm transforms the input DAG presented to RDS, allowing it to partition multiple output shaders, even for hardware which supports only a single render target.
- A pass-merging algorithm which allows partitions generated by RDS to be optimized for hardware with support

† {tfoley, mhouston, hanrahan}@graphics.stanford.edu

for multiple render targets. We derive two new algorithms for multipass partitioning, MRDS and MRDS$'$, that combine merging with RDS.
- Performance analysis demonstrating that MRDS produces more efficient partitions of fragment shaders than RDS. We show that even for shaders with only a single output value, partitions generated by MRDS can execute more efficiently on graphics hardware with MRT support.

2. Related Work

2.1. High-Level Shading Languages

Cook[Coo84] and Perlin[Per85] laid the groundwork for current shading languages. The Renderman shading language [HL90] is commonly used today for high-quality offline shading in software rendering systems. The Pixelflow shading system introduced pfman, a shading language for real-time rendering [OL98].

The Stanford Real-Time Shading Language (RTSL) system compiles a Renderman-like language for early programmable graphics hardware [PMTH01]. RTSL virtualizes "frequencies" of computation, allowing the compiler to select whether individual operations will be executed per-vertex or per-fragment.

NVIDIA's Cg [MGAK03] and Microsoft's HLSL [Mic03] are high-level languages for current graphics hardware that allow programs to be written for either the vertex or fragment processor and then compiled for a variety of hardware targets. While these languages do not encode any fixed resource limitations, the individual compiler targets have strict limits and will reject shaders that exceed them. Thus, programmers must write shader implementations for multiple hardware targets, manually breaking shaders that exceed resource limits into multiple passes.

The Sh system is a meta-compiler implemented in C++ for generating shaders at runtime and applying transformations to them [MQP02, MMT04]. The system does not explicitly specify resource limitations, although large shaders can currently fail to execute on graphics hardware.

The OpenGL Shading Language, GLSL, is a high-level languages for writing vertex and fragment processor shaders that makes it easy for programmable shaders to integrate with state from the fixed-function pipeline [KBR03]. The GLSL specification requires that implementations virtualize limits on number of instructions and temporary registers, but forces programmers to query and respect hardware-specific limitations on other resources.

2.2. Multipass Partitioning

Peercy, et al. map a subset of the Renderman language to a fixed-function OpenGL platform by abstracting the graphics hardware as a SIMD processor [POAU00]. Each rendering pass executes a single SIMD instruction, and a tree-matching approach is used to map shader computations to the small set of fixed-function operations available. While this technique generates good multipass partitions for fixed-function hardware, Proudfoot et al. demonstrate that tree-matching techniques are not sufficient for multipass partitioning on programmable hardware [PMTH01].

RTSL utilizes the RDS algorithm to virtualize fragment shading resource limits through multipass partitioning. As shown in [CNS*02], this allows for efficient partitioning of large shaders into multiple passes.

The Ashli system reads shaders in a number of high-level languages as input, including HLSL, GLSL, and Renderman, and generates low-level code to execute them on graphics hardware [ATI03a]. Ashli can generate multipass partitions for large shaders using RDS and provides the user with API calls to progressively render their scene. Ashli has demonstrated the effectiveness of RDS with multiple input languages.

Both the RTSL and Ashli systems are able to load balance shading computations between the vertex and fragment processors. Our implementation does not consider splitting shaders across the two shading units, only concentrating on partitioning shaders to run on the fragment processor.

2.3. Overview of the RDS Algorithm

Given a fragment shader represented as a DAG, the problem of generating a multipass partition is as follows: label a subset of the n DAG nodes as *splits*, intermediate values to be saved to texture memory, and then generate a partition of those splits into p rendering passes. Each pass then consists of shader code to generate its constituent splits as outputs, using texture fetches to restore the values of previously computed splits. Such a partition is *valid* if each of the passes can run on the target hardware and they can be ordered to preserve dependencies between splits. Among the many possible multipass partitions of a given shader, we wish to find one that is maximally efficient. However, the space of possible partitions is exponential, so the goal of multipass partitioning is to efficiently find a partition that is close to optimal.

The RDS algorithm combines top-down heuristic splitting, bottom-up greedy merging, and a limited search algorithm to mark nodes in a shader DAG as splits. We use the RDS algorithm as the base for our new techniques. RDS assumes that the target hardware can only output a single value per fragment in each rendering pass, and thus that the shader DAG has a single root. Along with this graph, RDS operates on its associated *dominator tree*. A DAG node x is a *dominator* of node y, written x dom y, if all paths from the root of the graph to y pass through x. The *immediate dominator* of node y is the unique $x \neq y$ such that x dom y and for all nodes $z \neq y$, z dom $y \Rightarrow z$ dom x. The dominator tree of a

DAG shares the node set of the graph, and connects each node as the direct child of its immediate dominator.

Some subset of the nodes in the DAG are *multiply-referenced* (MR) nodes, having more than one direct parent. The RDS algorithm is primarily concerned with whether these MR nodes should be marked as split locations. If a MR node is marked as a split, we subsequently incur the additional bandwidth costs of writing the value to texture memory and restoring that value in later passes. If the node is not marked as a split, we may incur additional computation costs from recomputing its value in multiple passes. Both of these additional costs can be eliminated if we can compute a MR node in the same pass as its immediate dominator, as all references to the MR node are then isolated to a single pass. The RDS algorithm attempts to eliminate save/recompute costs when possible and otherwise tries to chose the less expensive of the two options.

RDS uses two graph traversals to partition the DAG, a top-down subdivision over the dominator tree, and a bottom-up merge over the DAG. The subdivision traversal makes heuristic decisions about whether to save or recompute MR nodes, while the merge traversal greedily combines nodes with as many of their children as possible. These two traversals operate over the n nodes of the DAG and at each node use a low-level compiler to check the validity of a fixed number of subregions of the graph. Assuming that low-level compilation is a linear-time operation, this leads to an overall running time of $O(n^2)$. The graph traversals are wrapped in a limited search over the MR nodes of the DAG. The search algorithm forces successive MR nodes to be saved or recomputed (split, or unsplit) and compares the cost of partitions generated by the graph traversals under these two constraints. Each node is then fixed in whatever state led to the better partition and this information overrides the heuristic decisions made in the subdivide step. Using this limited search can increase the efficiency of generated partitions, but increases running times by a factor of n, making RDS an $O(n^3)$ algorithm.

3. The Algorithm

3.1. Multiple-Output Shaders

The RDS algorithm cannot operate on shaders with multiple output values since the dominator tree which drives the subdivision step is undefined when the shader DAG has multiple roots. A MR node m that can be reached from two different DAG roots (outputs) may have no immediate dominator, and thus would not be considered by the subdivide traversal of RDS.

We propose a simple solution to this problem that still allows us to take advantage of the information the dominator tree provides. Before applying RDS to a shader DAG with multiple root nodes, we insert a new node at the root of the DAG with operation *join* and having the shaders outputs as

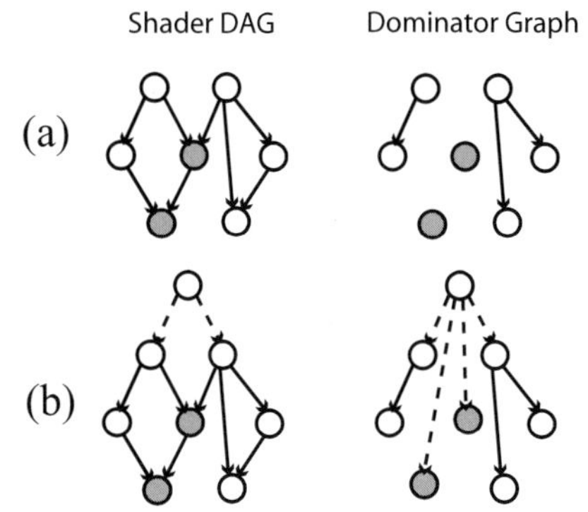

Figure 1: *(a) A multiple-output shader and its associated dominator graph. Note that the shaded intermediate nodes do not have immediate dominators, and the dominator graph is not a tree. (b) After adding a new root node to the DAG that joins the two outputs, the dominator graph is tree-structured.*

children. We define the join operation so that it compiles successfully if and only if all of its children are marked as splits. Figure 1 illustrates how this procedure is applied to a DAG with multiple roots. The left graphs represent shading DAGs and the right graphs are their associated dominator graphs. The addition of the join operation generate a single-rooted DAG from which a dominator tree can be derived. In this way we can partition multiple-output shaders into multiple single-output rendering passes while remaining transparent to the existing RDS algorithm.

The result of the above algorithm will be a set of splits, and for hardware that can only write a single output per pass we can simply assign each split to its own rendering pass to achieve good results. However, on graphics hardware that supports multiple render targets, we may be able to produce more efficient partitions.

3.2. Multiple-Output Hardware

Typically, the outputs of multiple-output shaders will not be disjoint, and thus intermediate results may be shared be tween outputs. The immediate dominator of such intermediate value nodes is the root join operation, and thus the dominator tree cannot help us to eliminate the save/recompute costs for such nodes. However, if we can write all of the outputs that depend on such a node m in the same pass that we use to compute m, we can avoid these additional costs.

This situation is not unique to shaders with multiple outputs. As Figure 2 demonstrates, even in shaders with only a single output, there may be sets of intermediate values that can be computed together much more efficiently than apart. We can extend the concept of dominance in a graph to describe this situation:

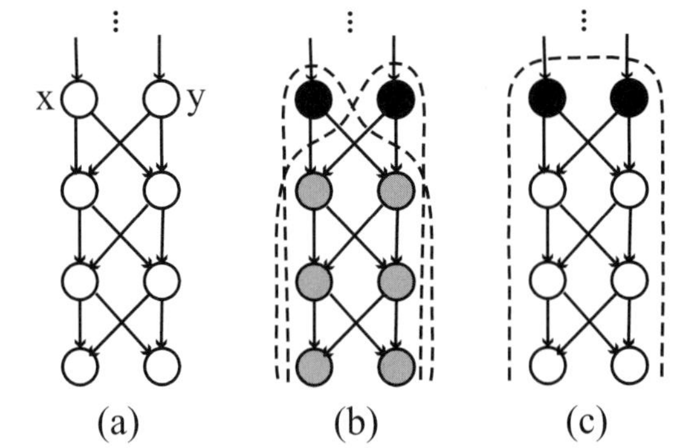

Figure 2: *A "ladder" configuration of intermediate value nodes. (a) The nodes x and y both depend on all previous intermediate values. (b) If x and y are marked as splits by a single-output partitioning algorithm, then the shaded nodes will have to be recomputed. (c) A MRT-aware algorithm can save both splits in a single pass and avoid the recomputation.*

Given a DAG node x and a set of nodes S, we say that x is a *dominating set* of x, written S dom x if and only if every path from a root of the DAG to node x passes through a node in S. The parents of a node x form a dominating set of x, and the set of shader outputs dominate every node in a shader DAG. Given this notation, we can generalize some of our previous statements about how dominance affects multipass partitioning: if a DAG node x is computed in a pass with output set S, where S dom x, then we can eliminate any save/recompute costs for x.

We can realize the benefits of this effect incrementally by merging the passes of an existing multipass partition. If an MR node n is being recomputed in many passes, and two of those passes are merged then some recomputation costs for m are eliminated. By repeatedly merging passes it may be possible to merge m with a dominating set and eliminate all of these costs. The following algorithm, MERGE-PASSES, uses greedy pass-merging to decrease recomputation costs in a multipass partition:

```
// input: T a set of splits as generated by RDS
MERGE-PASSES( T )
   // generate initial set of passes
   for each node n in T
      create a pass with output n and add it to the list of passes
   // create list of candidate merges
   for each pair (x, y) of nodes in T
      score ← TEST-MERGE( pass(x), pass(y) )
      if score ≥ 0 then
         add (score, x, y) to array A of potential merges
   sort A in decreasing order by score
   // attempt to execute merges from best to worst
   for each tuple (score, x, y) in A
      if pass(x) ≠ pass(y) then
         // score may have changed, so revalidate
         score ← TEST-MERGE( pass(x), pass(y) )
         if score ≥ 0 then
            EXECUTE-MERGE( pass(x), pass(y) )

// input: A, B passes (sets of nodes)
// output: a score measuring how "good" the merge is
TEST-MERGE( A, B )
   // if there is an ordering conflict, we cannot merge
   if ancestors(A) ∩ descendants(B) ≠ ∅
      ∨ descendants(A) ∩ ancestors(B) ≠ ∅
   then return −1
   M ← outputs(A) ∪ outputs(B)
   // use low-level compiler to check validity
   if pass-valid(M) then
      // score is how much the merge would improve partition cost
      cost_merged ← pass-cost(M)
      return cost(A) + cost(B) − cost_merged
   else return −1

// input: A, B passes (sets of nodes)
EXECUTE-MERGE( A, B )
   // create and initialize the merged pass
   create pass P
   outputs(P) ← outputs(A) ∪ outputs(B)
   cost(P) ← pass-cost(outputs(P))
   remove A, B from list of passes
   add P to list of passes
   // update membership information for nodes in this pass
   for each node n in outputs(P)
      pass(n) ← P
```

This algorithm operates on a list of splits and produces a list of passes. It begins by considering all possible pairwise merges between splits and determining whether they are valid. If there exists some split z such that z is an ancestor of one pass and a descendant of the other (i.e. z depends on the outputs of A and an output of B depends on z) we dismiss the merge as invalid. Otherwise we construct set M, the union of the output splits of passes A and B. If the elements of M cannot be generated in a single rendering pass on the target hardware, then the merge is invalid. For every valid merge we calculate a score measuring the improvement in cost of the merged pass over the two input passes, dismissing merges that yield negative scores. We iterate over the remaining potential merges in order of decreasing score and try to execute them. It is possible that earlier merges will have invalidated a potential merge or that it will no longer improve the overall score, so we re-check validity before executing any given merge.

The running time of this algorithm is dominated by the initial search for valid merges. For an n-node DAG and s splits, the search operates over s^2 pairs of splits and performs size-s set operations and an $O(n)$ compiler call. This yields an overall running time of $O(s^2(s+n))$ for MERGE-PASSES. In general, s is $O(n)$ in the number of DAG nodes, and thus the algorithm is $O(n^3)$, although in practice we expect s to be small relative to n.

If the MERGE-PASSES algorithm is run directly on the results of our multiple-output enabled RDS, then the result is an $O(n^3)$ algorithm for generating multipass partitions for MRT hardware. We call this algorithm MRDS, or Merging Recursive Dominator Split. We will demonstrate in the following section that MRDS can significantly improve the efficiency of multipass partitions generated by RDS.

However, it is possible that MRDS might produce poor partitions for certain shaders. Sets of splits that lead to expensive partitions when executed with one split per pass might yield highly optimized partitions after greedy merging. The search algorithm employed by RDS does not consider the potential for merging when evaluating the relative quality of partitions, and thus could make sub-optimal save/recompute decisions based on the incomplete information available to it. Thus, we introduce a second modified RDS algorithm, MRDS$'$.

The MRDS$'$ algorithm integrates greedy pass-merging directly into the search performed by RDS. When RDS evaluates the cost of a partition in order to make save/recompute decisions, we first apply pass-merging to collapse the candidate set of splits into a smaller number of passes. While this modification provides the RDS search with more accurate cost information, it comes with a penalty in asymptotic performance. The RDS algorithm wraps a linear search around an $O(n^2)$ graph traversal, and MRDS$'$ introduces the additional cost of $O(n^3)$ merging inside this search. Thus MRDS$'$ increases compile times by a factor of n over RDS, resulting in an $O(n^4)$ algorithm.

3.3. Implementation

We have implemented the RDS, MRDS and MRDS$'$ algorithms for multipass partitioning into the BrookGPU [BFH*04] system for stream computing on GPUs. Although the Brook language and the BrookGPU system are designed for stream computation rather than interactive shading, the implementation of our multipass partitioning algorithms are devoid of any concepts specific to streaming. The BrookGPU compiler transforms functions written in the Brook language into high-level fragment shaders, constructing a shader DAG from this representation. Our multipass partitioning algorithms operate on this DAG, using modularized compiler back-ends to encapsulate the validity testing and cost metrics required. In order to evaluate the effectiveness of MRDS and MRDS$'$ at generating multipass partitions for hardware with support for multiple render targets, we developed a back-end that generates DirectX 9 pixel shader code for the ATI Radeon 9800XT [Mic01, ATI03b]. This back-end uses the Microsoft HLSL compiler, fxc to perform validation and generate low-level code.

Whereas the RDS implementation used in RTSL operates over a DAG of low-level machine instructions, our system operates over a DAG of the high-level expression tree and relies on the external compiler to perform instruction selection. This makes implementing a back-end target much simpler as it need only pass high-level code to an external compiler. However, opportunities for optimization could be lost by operating on code before instruction selection. Our implementation will not split shader operations that are represented as primitive function calls, such as vector cross product, across passes even if they consist of multiple machine instructions. Furthermore, we generate DAG nodes for operations, such as swizzles and negation, which might be free on a particular hardware target.

3.4. Cost Metric

All of our multipass partitioning algorithms rely on a cost metric to provide predictions of shader performance on the target hardware. The original RDS implementation used a linear cost model that combined the total number of passes, arithmetic instructions and texture-fetch instructions in a given shader partition. In trying to model shader execution cost on our target platform, we have extended this model.

Chan et al. estimate the cost of each shader pass by measuring the time taken to render a single-pixel quad on the target hardware and convert this into units of GPU instruction cycles. The goal is to measure the cost of API overhead and setup to render a primitive, while making the cost of any per-fragment operations negligible [CNS*02]. However, because shader execution is typically asynchronous with API calls on the CPU, this overhead can be mitigated by running a shader on enough data so that GPU execution time dominates API overhead. We expect that in practice most applications of MRDS will apply to shaders being run on sufficiently large datasets to be GPU-limited rather than CPU-limited. Therefore we do not use the same procedure to derive per-pass overhead as Chan et al.

However, this is not to say that we do not account for perpass overhead. Experimentally we have found that the dominant overhead of each rendering pass is in writing shader results to off-chip framebuffer memory. We measured this cost by comparing execution times $t0$ when running i instructions over n fragments and $t1$ when running $i/2$ instructions over $2n$ fragments, always operating on enough data to be GPU-limited. If we fit a cost model of the form $cost = c_i \cdot i \cdot n + c_p \cdot n$ then we expect to find $c_p = (t1 - t0)/n$. We have experimentally measured c_p, the per-fragment pass overhead, to be equivalent to approximately 10 instructions on our target platform.

We must also take into account the total number of outputs being written in a given pass, as each output incurs an additional bandwidth penalty. We have found that shader output writes appear to mask the instruction execution time of small shaders. We have determined that single-output shaders with only a single instruction take the same amount of time to execute as those with 6 instructions, and that this number of instructions increases linearly as we increase the number of shader outputs.

We have measured the latency of a texture fetch on our target hardware to be equivalent to approximately 8 instructions for textures in float4 format. However, accounting for this delay with a linear term fails to take into account the potential for the hardware to schedule instructions to hide the latency of texture fetches. In the best case, the pipeline can completely hide texture latency and our execution time is $max(t_{textureFetch}, t_{instructionExecute})$, while in the worst case every texture fetch leads to a stall, and execution time is $t_{textureFetch} + t_{instructionExecute}$. Rather than try to inspect the texture-access behavior of a pass, we choose to model this term as a simple average of the best-case and worst-case forms.

The complete equation that we use to measure the cost c_{pass} of a shader pass consisting of i total instructions, t texture fetches and writing o output values is:

$$c_{pass} = c_p + max(c_o \cdot o, \frac{(i + c_t \cdot t) + max(i, c_t \cdot t)}{2})$$

where $c_p = 10$ is the per-fragment pass overhead, $c_o = 6$ is the number of instructions masked by each output write, and $c_t = 8$ is the latency of a texture fetch.

4. Results

In order to evaluate the quality of partitions generated by our new algorithms, we applied four different partitioning strategies to four different shaders written in the Brook language. The partitioning strategies were:

- Ideal - Represents a partitioning strategy that always puts all computation into a single rendering pass. The partitions generated by this strategy cannot be run on the target hardware, but are useful for comparison.
- RDS - The original RDS algorithm modified to accept multiple-output shaders.
- MRDS - The RDS strategy followed by our greedy pass-merging algorithm.
- MRDS′ - The RDS strategy with greedy pass-merging integrated into the RDS search step.

The applications were selected to provide a range of shader sizes, outputs, and uses. The individual applications are:

- Particle - A shader that advances a particle-system based cloth simulation. This shader uses two outputs to write the new position and velocity of each particle in the system. Each particle is constrained to up to 8 of its neighbors by spring forces. This application is representative of n-body dynamics simulations
- Fractal - A shader that computes a 40-iteration approximation to the Mandelbrot set. The shader is vectorized to process 4 adjacent points at once. This application was chosen to demonstrate that even shaders with a single output value may have strongly coupled intermediate values, and thus can benefit from hardware support for multiple outputs.
- Matrix - A shader that computes an 8 by 8 dense matrix multiply. Input and output matrices are packed with two float4 values per row. This shader shows how the merging algorithms, while not parameterized by the number of outputs supported in hardware, can partition shaders that use more than the number of supported outputs.
- Fire - A procedural volumetric fire shader [Ura02]. This shader uses 5 octaves of 4-dimensional Perlin noise, and demonstrates an extremely large shading computation with many levels of texture indirection.

Table 1 shows the results of partitioning and executing our applications under each of our partitioning strategies. These tables show the number of passes in each partition, along with its cost as given by our metric. The execution timing results represent the average per-fragment execution time, in nanoseconds, when running the partitioned shader on the graphics hardware. These values were generated by amortizing the time taken to shade a 1024 by 1024 pixel quad in the case of Particle and Fractal, and a 512 by 512 pixel quad in the case Matrix and Fire. All timing results were averaged over 1000 iterations. The compilation times represent the total time spent in the partitioning algorithm, including time spent waiting on the external compiler.

5. Discussion

Partitions generated by MRDS and MRDS′ outperform those generated by RDS for all of our shaders, including those that write only a single output. The relative increase in performance varies across the applications, and we will discuss these variations individually.

Although the Particle shader writes multiple output values, compiling for hardware with MRT support does not significantly increase its performance over a partition generated for single-output hardware. The majority of the instructions in the shader are used to calculate a sum of neighbor forces, and this running sum can be maintained with a single output per pass. The final integration of position and velocity is the only step that generates multiple values. Despite a 33% decrease in the number of passes, the merged partitions take only 5% less time to execute. This result indicates that per-pass overhead is not the dominant factor in shader performance.

The Fractal shader, despite writing only a single output value, shows a marked improvement in execution time when partitioned with MRDS or MRDS′ rather than RDS. The two merged partitions execute 47% more quickly than the RDS partition. We attribute this result to the fact that the Fractal shader maintains two live values throughout most of its body. These values are used in an iterative computation, with

Particle	Ideal	RDS	MRDS	MRDS′
Passes	1	6	4 (33%)	4 (33%)
Cost	276	449	389 (13%)	389 (13%)
ArithOps	170	210	208 (1%)	208 (1%)
TexOps	19	33	28 (15%)	28 (15%)
TimeExec	-	56	53 (5%)	53 (5%)
TimeComp	-	0.71	0.71	0.80

Fractal	Ideal	RDS	MRDS	MRDS′
Passes	1	10	4 (60%)	4 (60%)
Cost	225	666	324 (51%)	324 (51%)
ArithOps	205	355	211 (41%)	211 (41%)
TexOps	2	38	14 (63%)	14 (63%)
TimeExec	-	119	63 (47%)	63 (47%)
TimeComp	-	7.9	7.9	9.2

Matrix	Ideal	RDS	MRDS	MRDS′
Passes	1	32	10 (68%)	10 (68%)
Cost	401	2128	1028 (52%)	966 (55%)
ArithOps	208	656	326 (50%)	298 (55%)
TexOps	32	176	88 (50%)	82 (53%)
TimeExec	-	380	290 (24%)	279 (27%)
TimeComp	-	4.3	4.4	14

Fire	Ideal	RDS	MRDS	MRDS′
Passes	1	37	22 (36%)	22 (36%)
Cost	1630	3109	2799 (10%)	2722 (12%)
ArithOps	690	1334	1249 (6%)	1329 (0%)
TexOps	150	237	227 (4%)	213 (10%)
TimeExec	-	3172	2225 (30%)	1879 (41%)
TimeComp	-	78	78	244

Table 1: *Results of partitioning several shaders with RDS, MRDS and MRDS′. For each partition we list the number of passes, cost measure, number of arithmetic and texture operations,execution time (in nanoseconds per fragment), and compilation time (in minutes). Percentages measure relative improvement over the RDS partition.*

each successive value depending on both previous values, yielding the previously-described "ladder" configuration in the shader DAG. This ladder configuration forces the RDS partition to duplicate many calculations, leading to a 73% increase in the number of arithmetic operations over the Ideal partition. The merged partitions, on the other hand, introduce no additional arithmetic instructions beyond those needed to write intermediate outputs.

The Matrix shader shows modest improvements in execution time for both merged strategies. The MRDS and MRDS′ partitions are 24% and 27% more efficient, respectively, than the partition generated by RDS. As in the case of the Particle shader, a drastic decrease in the number of passes executed did not lead to a proportional increase in execution time. The Matrix shader also clearly demonstrates that our cost metric is not directly proportional to execution time. The cost of the RDS partition is approximately 2.2 times that of the MRDS′ partition, but takes only 36% more time to execute. However, it appears that relative differences in partition costs for a particular shader are predictive of relative performance.

The Fire shader also shows appreciable improvements of 30% and 40% over RDS when partitioned with MRDS and MRDS′. It is interesting to note that although the MRDS′ partition has a better score than the MRDS partition, it consists of the same number of passes, and actually executes more arithmetic instructions. Instead of minimizing instruction counts, the MRDS′ algorithm found a partition that reduced the total number of texture fetches at the expense of arithmetic operations. This is a sensible optimization for a bandwidth-limited shader, and our performance results show that these decisions yield a 16% increase in performance over MRDS.

Both MRDS and MRDS′ yield partitions of the Particle and Fractal shaders with the same overall cost and execution time, although the generated passes differ in the placement of a few instructions. For these relatively small shaders it appears that there exists a small set of nodes at which efficient splits can be made, and thus both MRDS and MRDS′ select similar points at which to split the DAG. For the larger Matrix and Fire shaders we find that the MRDS′ strategy generates partitions with slightly better performance. In the case of the Matrix shader, the MRDS′ partition has the same number of passes as the MRDS partition, but saves more intermediate values and thus avoids certain recomputation costs. Although the Fire shader saw modest gains in runtime performance from using MRDS′, we note that compilation with MRDS′ took over a factor of three times longer than with MRDS.

For all of our applications the RDS and MRDS compilations take a comparable amount of time. In the case of the Particle and Fractal shaders the MRDS′ algorithm increases these compile times by less than 17%. For these applications the number of splits is relatively small and the merge operations are quite efficient. In the case of the Matrix and Fire applications the number of splits is higher and the merges dominate the cost of the MRDS′ compiles. These results also demonstrate that the number of splits s is often much less than n, and the single merge step of MRDS is relatively inexpensive. We note that in every case between 94% and 97% of the compilation time was spent in the external compiler, fxc.

In general our results indicate that although the MRDS′ algorithm generates better partitions than MRDS for some shaders, the improvements in performance will often not be sufficient to justify the additional time complexity. However, the MRDS algorithm performs similarly to RDS, and improves performance on MRT hardware even for shaders that write only a single output value.

6. Conclusion and Future Work

We have described a set of modifications to the RDS algorithm that allow it to partition fragment shaders with multiple outputs, and to take advantage of hardware with support for multiple render targets. We have demonstrated that these modifications can allow partitions generated for hardware with multiple outputs to outperform partitions limited to single outputs, even for shaders producing only a single result. Although these modifications greatly expand the range of shaders to which multipass partitioning can be applied, a number of issues still remain.

While our implementation uses an external shader compiler to perform validation checks and measure resource usage, our results indicate that there is a clear benefit to integrating the MRDS algorithm for multipass partitioning into existing shader compilers. As the original RDS paper mentions, one way to improve the asymptotic bounds of the partitioning process is to utilize incremental compilation of shader passes. Using such a strategy we can effectively reduce the cost of compiler calls from $O(n)$ to $O(1)$.

It is useful to know how close the results of a heuristic algorithm, such as ours, are to optimality. The size of the search space involved, and the size of the shaders we use, make finding optimal partitions by brute force prohibitive. It may be possible to utilize directed search algorithms, such as A*, to find optimal partitions in far less time than a brute-force search. This would require the careful formulation of the state space to be searched, as well as the derivation of a good admissible heuristic to direct the search.

Our analysis of multipass partitioning assumes that a shader computation can be expressed as a single dataflow DAG, and thus does not handle loops, predication, or any data-dependent control flow. Future graphics hardware will support data-dependent branching in the fragment processor, and it is interesting to explore multipass partitioning algorithms that handle branching. It is possible that certain branching constructs will yield better performance when control flow is divided between the CPU and GPU, requiring partitioning algorithms to balance this load.

Although some hardware resource limits, such as instruction counts, have grown drastically with successive generations of programmable hardware, others, such as the number of available texture coordinate interpolants, have remained relatively low. Thus, shaders that use constrained resources will still require multipass partitioning. As instruction counts increase into the hundreds or thousands, a partitioning algorithm with running time polynomial in the number of shading instructions can be prohibitively expensive.

An interesting area for future research is to develop partitioning strategies that operate on dependency graphs of only the most constrained resources. This may be modeled as a constraint satisfaction problem, and limiting the search space in this way may drastically reduce compile times. Thinking of multipass partitioning in terms of resource allocation, this constraint search is closely related to register allocation and instruction scheduling for CPUs. Combining these two approaches, it may be possible to further improve the running time of multipass partitioning algorithms.

7. Acknowledgments

We would like to thank Eric Chan for the original RDS implementation from the Stanford Real-Time Shading Language system. Kekoa Proudfoot and Ren Ng provided invaluable feedback during the development of the MRDS algorithm. Kayvon Fatahalian provided the benchmarking results that drove our cost model.

This work was done on the Brook for GPUs system, which is supported by DARPA. Additional support has been provided by ATI, IBM, NVIDIA and SONY. The Brook programming language has been developed with support from Department of Energy (contract B527299-4), NNSA, under the ASCI Alliances program (contract LLL-B341491), the DARPA Smart Memories Project (contract MDA904-98-R-S855), and the DARPA Polymorphous Computing Architectures Project (contract F29601-00-2-0085).

References

[ATI03a] ATI: ASHLI - advanced shading language interface, 2003. http://www.ati.com/developer/ashli.html.

[ATI03b] ATI: Radeon 9800 technical specification, 2003. http://www.ati.com/products/radeon9800/radeon9800pro/specs.html.

[BFH*04] BUCK I., FOLEY T., HORN D., SUGERMAN J., FATAHALIAN K., HOUSTON M., HANRAHAN P.: Brook for GPUs: Stream computing on graphics hardware. In *Proceedings of ACM SIGGRAPH (to appear)* (2004).

[CNS*02] CHAN E., NG R., SEN P., PROUDFOOT K., HANRAHAN P.: Efficient partitioning of fragment shaders for multipass rendering on programmable graphics hardware. In *Proceedings of the conference on Graphics hardware 2002* (2002), Eurographics Association, pp. 69–78.

[Coo84] COOK R. L.: Shade trees. In *Proceedings of the 11th annual conference on Computer graphics and interactive techniques* (1984), ACM Press, pp. 223–231.

[HBSL03] HARRIS M. J., BAXTER W. V., SCHEUERMANN T., LASTRA A.: Simulation of cloud dynamics on graphics hardware. In *Proceedings of the ACM SIGGRAPH/EUROGRAPHICS conference on Graphics hardware* (2003), Eurographics Association, pp. 92–101.

[HL90] HANRAHAN P., LAWSON J.: A language for shading and lighting calculations. In *Proceedings of the 17th annual conference on Computer graphics and interactive techniques* (1990), ACM Press, pp. 289–298.

[KBR03] KESSENICH J., BALDWIN D., ROST R.: The OpenGL Shading Language, 2003. http://www.opengl.org/documentation/oglsl.html.

[MGAK03] MARK W. R., GLANVILLE R. S., AKELEY K., KILGARD M. J.: Cg: A system for programming graphics hardware in a c-like language. *ACM Transactions on Graphics 22*, 3 (July 2003), 896–907.

[Mic01] MICROSOFT: DirectX product web site, 2001. http://www.microsoft.com/directx/.

[Mic03] MICROSOFT: High-level shader language, 2003. http://msdn.microsoft.com/library/default.asp?url=/library/en-us/directx9_c/directx/graphics/reference/Shaders/HighLevelShaderLanguage.asp.

[MMT04] MCCOOL M. D., MOULE K., TOIT S. D.: Sh: Embedded metapgramming language, 2004. http://libsh.sourceforge.net/.

[MQP02] MCCOOL M. D., QIN Z., POPA T. S.: Shader metaprogramming. In *Proceedings of the ACM SIGGRAPH/EUROGRAPHICS conference on Graphics hardware* (2002), Eurographics Association, pp. 57–68. revised version.

[OL98] OLANO M., LASTRA A.: A shading language on graphics hardware: the pixelflow shading system. In *Proceedings of the 25th annual conference on Computer graphics and interactive techniques* (1998), ACM Press, pp. 159–168.

[PBMH02] PURCELL T. J., BUCK I., MARK W. R., HANRAHAN P.: Ray tracing on programmable graphics hardware. *ACM Transactions on Graphics 21*, 3 (July 2002), 703–712. ISSN 0730-0301 (Proceedings of ACM SIGGRAPH 2002).

[Per85] PERLIN K.: An image synthesizer. In *Proceedings of the 12th annual conference on Computer graphics and interactive techniques* (1985), ACM Press, pp. 287–296.

[PMTH01] PROUDFOOT K., MARK W. R., TZVETKOV S., HANRAHAN P.: A real-time procedural shading system for programmable graphics hardware. *ACM Transactions on Graphics* (August 2001).

[POAU00] PEERCY M. S., OLANO M., AIREY J., UNGAR P. J.: Interactive multi-pass programmable shading. In *Proceedings of the 27th annual conference on Computer graphics and interactive techniques* (2000), ACM Press/Addison-Wesley Publishing Co., pp. 425–432.

[Ura02] URALSKY Y.: Volumetric fire cg shader, 2002. http://www.cgshaders.org/shaders/show.php?id=39.

Graphics Hardware (2004)
T. Akenine-Möller, M. McCool (Editors)

Tile-Based Texture Mapping on Graphics Hardware

Li-Yi Wei

NVIDIA Corporation

Abstract

Texture mapping has been a fundamental feature for commodity graphics hardware. However, a key challenge for texture mapping is how to store and manage large textures on graphics processors. In this paper, we present a tile-based texture mapping algorithm by which we only have to physically store a small set of texture tiles instead of a large texture. Our algorithm generates an arbitrarily large and non-periodic virtual texture map from the small set of stored texture tiles. Because we only have to store a small set of tiles, it minimizes the storage requirement to a small constant, regardless of the size of the virtual texture. In addition, the tiles are generated and packed into a single texture map, so that the hardware filtering of this packed texture map corresponds directly to the filtering of the virtual texture. We implement our algorithm as a fragment program, and demonstrate performance on latest graphics processors.

Categories and Subject Descriptors (according to ACM CCS): I.3.1 [Computer Graphics]: Hardware Architecture: Graphics Processors; I.3.7 [Computer Graphics]: Three-Dimensional Graphics and Realism: Texture;

Keywords: Texture Mapping, Graphics Hardware, Texture Synthesis

1. Introduction

Texture mapping [Hec86] is a technique to represent surface details in computer rendered images without adding geometric complexity. Texture mapping has been a standard feature for recent consumer-level graphics hardware. The cost of texture mapping on graphics hardware includes the memory for storing textures, as well the bandwidth for transferring and accessing these textures. For applications that use large amounts of textures, the storage or bandwidth requirements may prohibit the graphics hardware from achieving real-time performance.

One possible solution to address these problems is texture compression [BAC96]. However, existing texture compression techniques are designed for general images and may achieve sub-optimal compression ratio for textures that contain repetitive patterns. For this kind of textures, texture synthesis algorithms [Wei02] usually provide better compression ratio. Unfortunately, previous texture synthesis algorithms are often too slow or too complex for real time applications on graphics hardware.

We address this problem by implementing a large virtual texture as a stochastic tiling of a small set of texture image tiles [CSHD03]. Our technique functions transparently as a traditional texture mipmap and provides the illusion of the availability of a large, non-periodic virtual texture while consuming only a small amount of texture memory. We implement our technique as a fragment program and demonstrate performance on current generation programmable graphics hardware.

2. Previous Work

A variety of techniques have been proposed to reduce the texture storage requirements. We classify these techniques into three categories : Texture Compression, Texture Synthesis, and Texture Tiling.

Texture Compression : One way to reduce texture memory consumption is to compress the textures. [BAC96] presented a texture compression scheme based on Vector Quantization (VQ). This technique allows fast texture decoding while achieving a compression ratio up to 35:1, but may require the entire texture VQ codebook to be stored on-chip. Most current graphics hardware vendors adopt a variation of VQ compression, termed S3TC [S3 98], that does not re-

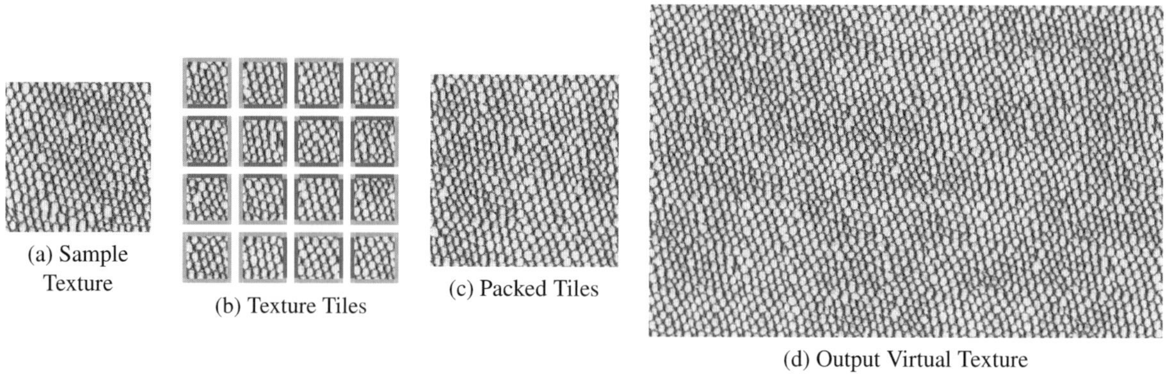

Figure 1: Overview of our system. Given a sample texture (a), we first construct a set of texture tiles [CSHD03] (b) so that adjacent tiles have continuous pattern across shared edges. We pack the tiles into a single texture map (c) and store it on a graphics processor. This texture map can then be sampled and filtered via a fragment program to support an arbitrarily large virtual texture (d) without physical storage.

quire the on-chip storage of entire codebooks. Instead, the textures are compressed by encoding each 4 by 4 pixel tile into a 64 bit data chunk. Unfortunately, the compression ratio of S3TC is only about 6:1. [Fen03] developed a compression scheme that could achieve very low bit rates around 4 or 2 bits per texel. However, the compressed data size is still linearly proportional to the original texture size.

[KE02] proposed a technique to compress textures adaptively into blocks of different resolutions. Unlike Vector Quantization where the compressed data is linearly proportional to the original data size, their technique employees adaptive block size and may achieve higher compression ratio for sparse textures. However, the technique as presented does not fully support texture filtering and mipmapping.

A related idea to texture compression is texture caching on graphics hardware [HG97, TMJ98]. These techniques can help reduce texture bandwidth, but not texture memory consumption.

Texture Synthesis : Texture compression techniques as presented above are designed for general texture images. However, for a specific class of textures that are composed of repeating elements, texture synthesis algorithms can offer higher compression rates. Techniques such as [EF01, Wei02, ZG02, KSE*03] only require a small texture sample to generate arbitrarily large results, so in some sense the compression ratio can approach infinity for really large textures. Unfortunately, these techniques are often too slow or too complex for graphics hardware implementation. Instead, they would pre-compute large textures on CPUs and consume a comparably large amount of texture storage on graphics chips.

[PFH00, SCA02] proposed a variant of these approaches by synthesizing texture coordinates rather than texture pixels on mesh surfaces. Although the result mesh can be rendered using the original small texture sample, these approaches require the application programs to pre-compute texture coordinates. In addition the computation is only valid for a specific combination of texture and mesh model.

Procedural texturing algorithms [Per02, Har01] allow textures to be computed efficiently on GPUs, but can only reproduce a limited class of textures such as marble, wood, and clouds. [MGW01, WWT*03] presented variations of texture mapping on graphics hardware, but they did not address issues with large textures.

Texture Tiling : An alternative approach is to use texture synthesis to pre-compute a small set of tiles and use these tiles to generate a large, non-periodic texture at run time. [LN03] procedurally combines texture tiles via indirection to generate large virtual textures. The technique requires only a small amount of texture memory, but as the authors discussed, texture filtering and mipmapping are inherently incorrect with indirection. [CSHD03] introduced Wang Tiles as texturing primitive and discussed methods to construct these tiles, but did not address the implementation issues on graphics hardware.

Computation on GPUs : With the advance of recent programmable graphics hardwares (GPU), many techniques have been proposed to map general computation kernels on GPUs to solve different rendering problems [BFGS03, KW03, Har03]. Our approach also utilizes the programmability of GPUs, but we present a new algorithm to resolve an existing hardware problem, texture storage and bandwidth, rather than directly porting an existing rendering algorithm to run on GPUs.

3. Overview

Our goal is to design a new algorithm that combines the advantages of previous approaches. The algorithm should be general enough to handle textures with repeating patterns. It should be simple, efficient, and consumes minimum amount of texture memory for graphics hardware implementation. It should support texture filtering and mipmapping. It should also be non-intrusive to application programs and does not require modifications on input geometry or texture coordinates.

We attempt to achieve these goals by using Wang Tiles [CSHD03] to generate large virtual textures directly on GPUs. Wang Tiles are square tiles in which each tile edge is assigned a color. A valid tiling requires all shared edges between tiles to have matching colors. When a set of Wang Tiles are filled with texture patterns that are continuous across matching tile edges, a valid tiling from such a set can produce an arbitrarily large texture without pattern discontinuity. [CSHD03] discussed techniques to construct such tile sets from sample images, and presented a sequential algorithm for non-periodic tiling.

We present a new texture mapping algorithm based on Wang Tiles. Our algorithm assembles these tiles on the fly to build a large virtual texture rather than physical storage. As a preprocess step, we pack these tiles into a single texture map, and our packing scheme ensures correct mipmap filtering when a texel is fetched from this tile pack. During run time, for each texture request (s, t), we first determine which input tile it lands at based on the position of (s, t) within the output texture. We then compute the relative offset of (s, t) within that input tile, and fetch the corresponding texel from the input pack. Since out packing scheme supports correct mipmap filtering, the fetched input texel will appear to be the same as a texel fetched from a large, physically stored texture map. An overview of our system is illustrated in Figure 1.

Our contributions are as follows. We present a new tiling technique to allow random access at arbitrary tiles. We propose a new method to pack these tiles into texture memory to support mipmap filtering, and demonstrate an efficient implementation of our algorithm on graphics hardware.

4. Tile-Based Texture Mapping

The goal of texture mapping is to assign a texel Tex(s, t) for each screen pixel with texture coordinate (s, t). Current graphic hardware stores the entire texture in the memory hierarchy, and implements Tex(s, t) by fetching and filtering relevant texels. This approach works well for small textures that can fit into on-chip texture cache or off-chip texture memory, but has difficulty storing and accessing large textures that cannot fit into available texture memory.

We present an alternative approach to Tex(s, t), TileTex(s, t), that does not require large texture storage. TileTex(s, t) provides the same interface as traditional Tex(s, t), supporting filtering and mipmapping for large textures. Internally, TileTex(s, t) represents a large texture as a small set of Wang Tiles, and assembles the tiles into a large virtual texture on the fly for answering texture

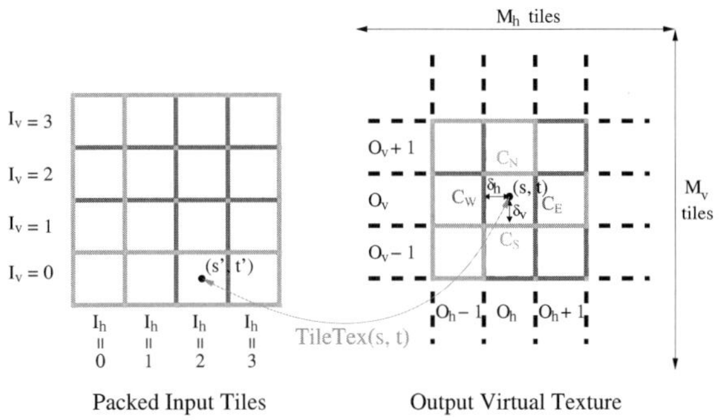

Figure 2: Graphical illustration of symbols used in our algorithm.

Inputs
- **S**: the sample texture
- $T_h \times T_v$: the individual tile size in pixels
- K_h, K_v : number of horizontal and vertical edge colors
- $M_h \times M_v$: output texture size in # of tiles

Preprocessing
- Construct the Wang tiles from **S** // *Section 4.1*
- Pack the tiles into TilePack // *Section 4.4*
- $(P_h, P_v) \leftarrow (K_v \times K_v, K_h \times K_h)$ // *TilePack size in # of tiles*
- **if** need CornerPack
  - Build a dual corner packing CornerPack // *Appendix B*
  - // *CornerPack has size $P_v \times P_h$*

Runtime $u_f \leftarrow$ TileTex(s, t)
- $(O_h, O_v) \leftarrow$ floor((s, t) $\times (M_h, M_v)$) $\%(M_h, M_v)$
- $(C_S, C_E, C_N, C_W) \leftarrow$ EdgeHash(O_h, O_v) // *Equation 2*
- // *The above 2 steps compute TileMap(s, t), as described in Section 4..*
- $(I_h, I_v) \leftarrow$ TileIndex(C_S, C_E, C_N, C_W) // *Equation 7*
- $(\delta_h, \delta_v) \leftarrow$ fraction((s, t) $\times (M_h, M_v)$)
- $u_e \leftarrow \mathrm{Tex}(\frac{I_h+\delta_h}{P_h}, \frac{I_v+\delta_v}{P_v})$ // *texel fetch from TilePack*
- **if** no CornerPack
  - **return** u_e
- **else**
  - $(Q_h, Q_v) \leftarrow \mathrm{floor}\Big((s, t) \times (M_h, M_v) - (\frac{1}{2}, \frac{1}{2})\Big) \%(M_h, M_v)$
  - $(C_B, C_R, C_T, C_L) \leftarrow$ CornerHash(Q_h, Q_v) // *Equation 12*
  - $(J_h, J_v) \leftarrow$ CornerIndex(C_B, C_R, C_T, C_L) // *Equation 10*
  - $(\varepsilon_h, \varepsilon_v) \leftarrow \mathrm{fraction}\Big((s, t) \times (M_h, M_v) - (\frac{1}{2}, \frac{1}{2})\Big) + (\frac{1}{2}, \frac{1}{2})$
  - $u_c \leftarrow \mathrm{Tex}(\frac{J_h+\varepsilon_h}{P_v}, \frac{J_v+\varepsilon_v}{P_h})$ // *texel fetch from CornerPack*
  - **return** Interpolate(u_e, u_c)

Table 1: Summary of our algorithm. The graphical illustration of some symbols is shown in Figure 2, and the Cg code implementation of **Runtime** is shown in Appendix C.

requests. Note that our algorithm is implemented entirely on the texture space and does not require applications to modify the input geometry or texture coordinates.

We now describe our tile-based texture mapping algorithm TileTex(s, t).

4.1. Texture Tile Construction from Input Sample

The inputs of our algorithm consist of a sample texture **S**, the individual tile size $T_h \times T_v$, the number of horizontal colors K_h for south (S) and north (N) edges, and the number of vertical colors K_v for east (E) and west (W) edges. These input parameters have the same meanings as in [CSHD03] and we directly adopt their approaches to build the Wang Tile set.

However, there remains an important difference between [CSHD03] and our approach in required number of tiles. [CSHD03] uses a sequential tiling method and requires only $K_h \times K_v \times 2$ tiles to support all possible NW color combinations for non-periodically tiling. Since our approach supports non-sequential, random tile access, we require $K_h{}^2 \times K_v{}^2$ tiles to support all possible edge color combinations. This is a disadvantage of our approach compared to [CSHD03]. But we haven't found it to be a major issue since in practice K_h and K_v are usually small (2 is sufficient for non-periodic tiling). In addition, a tile set with complete color combinations has further advantages such as the ease to support continuous packing for texture filtering, as we shall see later.

4.2. Texture Tile Mapping for Random Access

Once we build the tile set, we can tile them to generate arbitrarily large textures. To ensure continuous texture pattern, the tiling needs to be done so that all shared tile edges have identical colors. One method to achieve such a tiling is by putting the tiles sequentially from north to south and west to east, as shown in [CSHD03]. However, sequential tiling has a slight disadvantage that if only the SE tile is accessed, we still need to compute the entire tiling to make sure we end up with the same tile configuration at the SE corner. This can cause efficiency issues for graphics hardware implementation.

One possible solution is to pre-compute the tile mapping and store the results in an index texture map. However, the index texture can still be too large for a large output texture composed of small tiles.

Our approach is to compute the tile mapping on the fly rather than storing a pre-computed index map. For rendering consistency, we require our mapping computation to be *random-accessible*, meaning that the computation of each tile can be performed independently while ensuring all shared tile edges have identical colors. This avoids the sequential dependency problem and results in much faster computation. Our tile mapping computation, TileMap(s, t), is as follows:

- For an input texture coordinate (s, t), compute which output tile it lands on. Assume the output texture contains M_h horizontal tiles and M_v tiles. Then the output tile index (O_h, O_v) can be computed by:

$$(O_h, O_v) = \text{floor}((s, t) \times (M_h, M_v)) \,\%(M_h, M_v) \quad (1)$$

  where % is modular operator. We use % for toroidal boundary handling but it can easily be extended for other boundary wrapping modes.
- Use (O_h, O_v) to hash the 4 edge colors C_S, C_E, C_N, C_W, of the output tile, as follows.

$$\begin{aligned} C_S &= \text{hash}[\text{hash}[O_h] + O_v] \,\%K_h \\ C_E &= \text{hash}[(O_h + 1) \,\%M_h + \text{hash}[2\times O_v]] \,\%K_v \\ C_N &= \text{hash}[\text{hash}[O_h] + (O_v + 1) \,\%M_v] \,\%K_h \\ C_W &= \text{hash}[O_h + \text{hash}[2\times O_v]] \,\%K_v \end{aligned} \quad (2)$$

  where hash[] is a 1D hash function which we currently implement via a permutation table as described in [Per02]. We have found it sufficient to use a hash[] table with $\max(M_h, M_v)$ entries. Since M_h and M_v are often small the caching and storage of hash[] is usually not an issue.

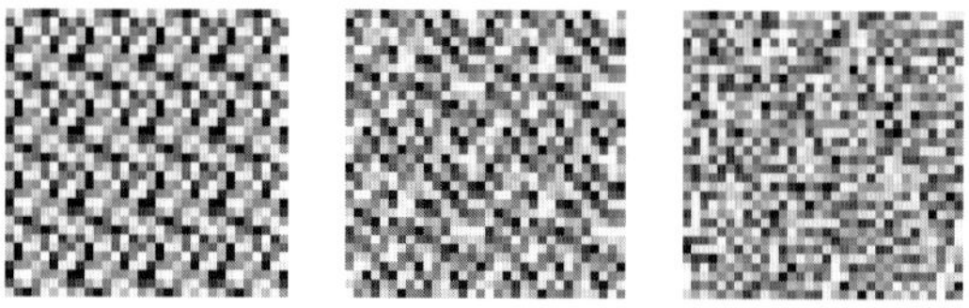

(a) 8 entries (a) 16 entries (a) 32 entries

Figure 3: Effect of hash table size on Equation 2. The image sizes are 32×32, and each pixel color encodes a unique combination of C_S, C_E, C_N, and C_W.

  Our mapping computation is inspired by a discrete chaos map called "Cat-Map" [XGS00]. In Cat-Map, a pair of integers (x, y) is mapped to a new location (x', y') by

$$\begin{aligned} x' &= x + y \\ y' &= x + 2 \times y \end{aligned} \quad (3)$$

  Cat-Map helps to break diagonal symmetry and increase randomness in tile mapping computation. In the equations above, C_S and C_W are adopted directly from Cat-Map, and C_N and C_E are derived from C_S and C_W to preserve edge color consistency. For example, consider tile (O_h, O_v) and the tile on the E side of it $(O_h + 1, O_v)$. The computation above ensures that C_E of (O_h, O_v) is the same as C_W of $(O_h + 1, O_v)$. Similar arguments can be applied to the computation of C_S and C_N.
- Select the input tile that has the corresponding 4 edge colors. Since our tile set has complete color combinations, this is always possible.

In summary, given a texel request at texture coordinate

(s, t), the above steps compute TileMap(s, t), specifying which input tile (s, t) lands on. The input tile is designated by the edge colors C_S, C_E, C_N, and C_W.

4.3. Texture Tile Access and Filtering

After knowing which input tile $I(C_S, C_E, C_N, C_W)$ the texture request (s, t) locates on, we need to figure out the relative position of (s, t) inside the tile to fetch correct texels. This relative position (δ_h, δ_v) can be computed by

$$(\delta_h, \delta_v) = \text{fraction}((s, t) \times (M_h, M_v)) \qquad (4)$$

where fraction() takes the fractional part of a floating point value. For nearest-neighbor texture sampling, the final texel value TileTex(s, t) will come from the texel nearest (δ_h, δ_v) within the corresponding input tile. For other sampling modes such as bilinear or anisotropic filtering, the issue will be more complicated. If (δ_h, δ_v) is sufficiently far away from the tile border, then it is adequate to perform the desired filtering operations within that tile. However, if (δ_h, δ_v) is close to the border, then correct filtering will involve texels from adjacent tiles in the output texture. Since the output texture is virtual, this filtering operation cannot be directly performed on graphics hardware.

There are several possible way to support filtering when (δ_h, δ_v) is close to the border. One possible solution is to set the texture filtering mode to be point sampling, and use a fragment shader to perform texture filtering. This is the most flexible method since we can access any texels across tile boundaries, but it can be slow since we need to request many texture samples (8 samples for trilinear mipmap filtering and possibly more for anisotropic filtering). Usually it is much more efficient to let the texture unit perform texture filtering since texture reads are relatively slow compared to arithmetic operations inside a fragment shader.

However, since hardware texture filtering is performed in the stored texture map, we have to find a way to store the set of Wang Tiles to support correct filtering. There are several possible ways to store the tiles. One solution is to store each tile as an individual texture. Since most hardware has a limited number of bind-able textures we can run out of texture IDs quickly. This approach also does not handle the tile boundary condition correctly. Another solution is to pack all the tiles into one single texture map. Assume the input packing has P_h horizontal tiles and P_v vertical tiles and the desired input tile is located as the $(I_h, I_v)^{th}$ tile in this packing, then we can fetch the correct texel for (s, t) with offset (δ_h, δ_v) as follows:

$$\text{TileTex}(s, t) = \text{Tex}\left(\frac{I_h + \delta_h}{P_h}, \frac{I_v + \delta_v}{P_v}\right) \qquad (5)$$

The challenge of this scheme is to make sure that texture filtering across output tile boundaries can be performed as filtering in the input tile pack. To achieve this goal, we present a packing scheme that leverages existing hardware texture filtering and mipmapping capability, and describe how to efficiently compute (I_h, I_v) from tile edge colors (C_S, C_E, C_N, C_W) in this packing scheme.

4.4. Texture Tile Packing and Corner Handling

In packing texture tiles into a single texture map we would like to achieve three goals. First, to utilize minimum amount of texture memory, each tile should appear only once in the packed texture. Second, to avoid filtering artifacts across texture tile borders, adjacent tiles in the packed texture should have matching edge colors. Third, the packing scheme should support efficient indexing. That is, given an edge color description of the input tile $I(C_S, C_E, C_N, C_W)$, the packing scheme should efficiently compute the horizontal and vertical indices (I_h, I_v) of $I(C_S, C_E, C_N, C_W)$ inside the packed texture.

We present a new tile packing scheme, TilePack, that attempts to satisfy the above requirements. For clarify, we first describe how to pack tiles in 1D. We then describe how to extend it to 2D.

Assuming we have a set of Wang Tiles that has K_v vertical edge colors but only 1 horizontal edge color. Our goal is to pack these tiles in a horizontal row so that adjacent tiles have matching vertical edge color (including the left-most and right-most tiles since the packed texture needs to be tileable). We encode all the vertical edge colors into integers in the range $[0..K_v - 1]$. Given a tile with west and east edge colors (e_1, e_2), we use the following mapping to determine the horizontal position of that tile inside the packed 1D texture: (The thought process for deriving this formula can be found in Appendix A.)

$$\text{TileIndex}_{1D}(e_1, e_2) = \begin{cases} 0, & e_1 = e_2 = 0 \\ {e_1}^2 + 2 \times e_2 - 1, & e_1 > e_2 > 0 \\ 2 \times e_1 + {e_2}^2, & e_2 > e_1 \geq 0 \\ (e_1 + 1)^2 - 2, & e_1 = e_2 > 0 \\ (e_1 + 1)^2 - 1, & e_1 > e_2 = 0 \end{cases} \qquad (6)$$

TileIndex_{1D} satisfies the three requirements listed above. It packs the $K_v \times K_v$ tiles into a horizontal 1D texture map without using any tile more than once. It ensures all shared tile edges have identical colors. In addition, it is efficient to compute, involving only multiplications, additions, and subtractions. An example of 1D packing with $K_v = 3$ is shown in Figure 4.

We can implement 2D TileIndex by TileIndex_{1D} on horizontal and vertical colors orthogonally, as shown in Equation 7. Given a tile with edge colors (C_S, C_E, C_N, C_W), the horizontal index I_h of the tile is computed by $\text{TileIndex}_{1D}(C_W, C_E)$, while the vertical index I_v of the tile is computed by $\text{TileIndex}_{1D}(C_S, C_N)$. Since TileIndex_{1D} satisfies the three requirements in 1D, and (I_h, I_v) are computed orthogonally, it can be easily seen that TileIndex meets the

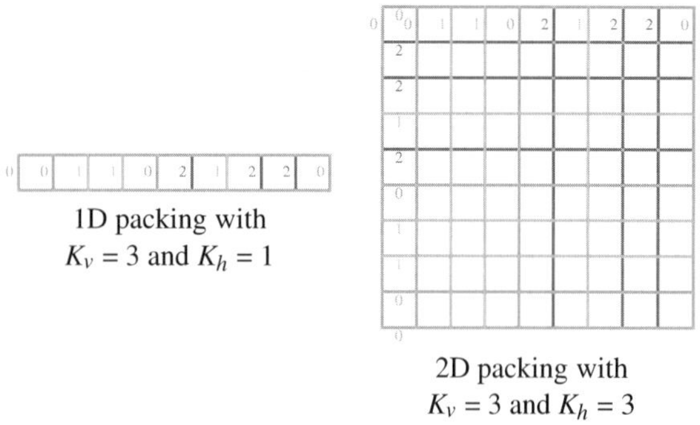

Figure 4: Tile packing examples. The illustration on the left shows a 1D tile packing with 3 vertical colors, and the one on the right shows a 2D tile packing with 3 vertical and 3 horizontal colors.

three requirements in 2D. An example of 2D packing with $K_h = 3$ and $K_v = 3$ is shown in Figure 4.

$$(I_h, I_v) = \text{TileIndex}(C_S, C_E, C_N, C_W) = (\text{TileIndex}_{1D}(C_W, C_E), \text{TileIndex}_{1D}(C_S, C_N)) \quad (7)$$

One of the primary advantages of our tile packing scheme is that it allows us to perform texture filtering directly via the hardware texturing unit. Since adjacent tiles in the packing share matching edge colors, there will be no pattern discontinuity when performing bilinear filtering across tile edges. However, bilinear filtering can be theoretically incorrect across tile corners. One solution to this problem is to simply ignore it. We have found that this corner artifact noticeable only when there is a distinctive object or pattern placed near a tile corner. For textures with uniform repetitive patterns, this is usually not an issue.

Another solution is to add an additional tile packing with all possible corner combinations. This solution would allow correct filtering across tile corners, but consumes extra amount of texture memory and shader computations. The derivation of this corner packing is analogous to the tile packing above, and we refer the reader to Appendix B for more details.

5. Results and Discussion

We have implemented our algorithm as a Cg fragment program [MGAK03] as shown in Appendix C. Our approach supports correct mipmap filtering across tile boundaries, as demonstrated in Figure 5. The performance of our implementation measured on a 350 MHz Geforce FX 5600 graphics processor [NVI03] is 2.7 million trilinearly-filtered texture samples per second without any hand optimized assembly code. This is slower than standard texture mapping which achieved at least 60 million texture samples per second on the same graphics card. However, the speed and storage requirement of our algorithm remain roughly constant regardless of the size of the output virtual texture. In comparison, the speed and storage requirement of traditional texture mapping degrade with increasing texture sizes, and most graphics chips impose an upper limit on the available texture size.†

At small texture sizes, our approach is slower than traditional texture mapping due to the run-time computation of TileMap(s, t) (Section 4.2), which would consume 8 accesses to the hash[] texture per request. For speedup, we have provided an alternative implementation that precomputes the TileMap(s, t) into a separate texture map. This approach has a better performance with more than 20 million trilinearly-filtered samples per second, with the cost of an extra memory for storing the TileMap(s, t). A better solution would be to utilize a built-in hash generator in the hardware, but unfortunately this feature is not yet available in current generation graphics chips. With such a hardware hash generator, we believe our first implementation would achieve comparable performance with respect to traditional texture mapping at small texture sizes.

Another limitation of our approach is that our tile filtering scheme is theoretically incorrect at lower mipmap levels where the filtering can access texels from more than 2 tiles. In practice, this is usually not an issue for textures with repeating patterns since the patterns reduce to homogeneous void at lower resolutions anyway. In addition this is an inherent limitation for any tile-based texturing algorithms unless you are willing to precompute and store lower mipmap levels for the output virtual texture.

Our approach can be implemented as part of the driver and exposed as an extension of standard API calls. For example, the **Preprocessing** part in Table 1 can be implemented as a special case of glTexImage2D, and the **Runtime** part can be implemented by letting the driver replacing any texture access code to a tile-based texture map with our fragment program. The result would be a transparent driver implementation that allows application programmers to access our tile-based texture map just like a traditional texture map.

6. Conclusions and Future Work

We present a tile-based texture mapping algorithm that allows the creation of a large, non-periodic, virtual texture map from a small set of pre-computed tiles. We propose a novel packing scheme that allows correct texture filtering across tile boundaries, as well a run time algorithm for sampling individual texels on the virtual texture map that can be directly implemented as a fragment program on current generation graphics processors.

† The maximum texture size is 4K × 4K pixels on a Geforce FX 5600 GPU.

(a) Tiles (b) Result with correct packing (c) Result with incorrect packing

Figure 5: Filtering artifact example. When the input tile set (a) is correctly packed into a single texture map via our method, the output texture, when rendered with bilinear mipmap filtering, exhibits no border artifact as shown in (b). However, if the input tiles are not properly packed, the rendering result may exhibit noticeable boundary artifacts as shown in (c).

There are several possible directions for future work. Our hash operation in Equation 2 relies on an input hash table, which consumes extra memory and additional texture fetches. This limitation can be overcome by replacing the hash[] table in Equation 2 with a well behaved hash algorithm such as MD5 implemented either as a fragment program or as part of the graphics hardware. However, since most hash functions require bitwise operations, a fragment program implementation would require integer and bitwise arithmetic beyond the floating-point-only instructions in current generation graphics chips. Another potential future work is to extend our approach to 3D textures. This would involve more shader program computations to sample 3D tiles, but could provide even higher compression ratio compared to 2D textures.

Acknowledgements

I would like to thank Wei-Chao Chen for his insights and discussions on an early draft of this paper, the anonymous reviewers for their invaluable suggestions, and my NVIDIA colleagues for their support and encouragement.

References

[BAC96] BEERS A. C., AGRAWALA M., CHADDHA N.: Rendering from compressed textures. In *Proceedings of SIGGRAPH 1996* (1996), pp. 373–378.

[BFGS03] BOLZ J., FARMER I., GRINSPUN E., SCHRÖDER P.: Sparse matrix solvers on the gpu: Conjugate gradients and multigrid. *ACM Transactions on Graphics 22*, 3 (July 2003), 917–924.

[CSHD03] COHEN M. F., SHADE J., HILLER S., DEUSSEN O.: Wang tiles for image and texture generation. *ACM Transactions on Graphics 22*, 3 (July 2003), 287–294.

[EF01] EFROS A. A., FREEMAN W. T.: Image quilting for texture synthesis and transfer. In *Proceedings of ACM SIGGRAPH 2001* (Aug. 2001), Computer Graphics Proceedings, Annual Conference Series, pp. 341–346.

[Fen03] FENNEY S.: Texture compression using low-frequency signal modulation. In *Graphics Hardware 2003* (July 2003), pp. 84–91.

[Har01] HART J. C.: Perlin noise pixel shaders. In *Proceedings of the ACM SIGGRAPH/EUROGRAPHICS workshop on Graphics hardware* (2001), pp. 87–94.

[Har03] HARRIS M.: General-purpose computation on gpus, 2003. http://www.gpgpu.org/.

[Hec86] HECKBERT P. S.: Survey of texture mapping. *IEEE Computer Graphics & Applications 6*, 11 (November 1986), 56–67.

[HG97] HAKURA Z. S., GUPTA A.: The design and analysis of a cache architecture for texture mapping. *24th International Symposium on Computer Architecture* (1997).

[KE02] KRAUS M., ERTL T.: Adaptive Texture Maps. In *Proc. SIGGRAPH/EG Graphics Hardware Workshop '02* (2002), pp. 7–15.

[KSE*03] KWATRA V., SCHÖDL A., ESSA I., TURK G., BOBICK A.: Graphcut textures: Image and video synthesis using graph cuts. *ACM Transactions on Graphics 22*, 3 (July 2003), 277–286.

[KW03] KRÜGER J., WESTERMANN R.: Linear algebra operators for gpu implementation of numerical algorithms. *ACM Transactions on Graphics 22*, 3 (July 2003), 908–916.

[LN03] LEFEBVRE S., NEYRET F.: Pattern based procedural textures. In *Proceedings of SIGGRAPH 2003 Symposium on Interactive 3D Graphics* (2003).

[MGAK03] MARK W. R., GLANVILLE R. S., AKELEY K., KILGARD M. J.: Cg: A system for programming graphics hardware in a c-like language. *ACM Transactions on Graphics 22*, 3 (July 2003), 896–907.

[MGW01] MALZBENDER T., GELB D., WOLTERS H.: Polynomial texture maps. In *Proceedings of ACM SIGGRAPH 2001* (Aug. 2001), Computer Graphics Proceedings, Annual Conference Series, pp. 519–528.

[NVI03] NVIDIA CORPORATION: GeForce FX Graphics Chip Series, 2003. http://www.nvidia.com/.

[Per02] PERLIN K.: Improving noise. *ACM Transactions on Graphics 21*, 3 (July 2002), 681–682.

[PFH00] PRAUN E., FINKELSTEIN A., HOPPE H.: Lapped textures. In *Proceedings of ACM SIGGRAPH 2000* (July

2000), Computer Graphics Proceedings, Annual Conference Series, pp. 465–470.

[S3 98] S3 CORPORATION: S3TC Texture Compression Standard, 1998.

[SCA02] SOLER C., CANI M.-P., ANGELIDIS A.: Hierarchical pattern mapping. *ACM Transactions on Graphics 21*, 3 (July 2002), 673–680.

[TMJ98] TANNER C. C., MIGDAL C. J., JONES M. T.: The clipmap: A virtual mipmap. In *Proceedings of SIGGRAPH 98* (July 1998), pp. 151–158.

[Wei02] WEI L.-Y.: *Texture Synthesis by Fixed Neighborhood Searching*. PhD thesis, Stanford University, 2002.

[WWT*03] WANG L., WANG X., TONG X., LIN S., HU S., GUO B., SHUM H.-Y.: View-dependent displacement mapping. *ACM Transactions on Graphics 22*, 3 (July 2003), 334–339.

[XGS00] XU Y.-Q., GUO B., SHUM H.: *Chaos Mosaic: Fast and Memory Efficient Texture Synthesis*. Tech. Rep. MSR-TR-2000-32, Microsoft Research, 2000.

[ZG02] ZELINKA S., GARLAND M.: Towards real-time texture synthesis with the jump map. In *Rendering Techniques 2002: 13th Eurographics Workshop on Rendering* (June 2002), pp. 99–104.

Appendix A: Derivation of Equation 6

Assume we have $N+1$ tile edges labeled from 0 to N. Our goal is to connect them in a tour so that each pair of edge combination appears exactly once. (This is another way to state the requirements listed in Section 4.4.) Mathematically, we need to find the Euler circuit for a directed, complete graph whose nodes are labeled from 0 to N. Denote the Euler circuit of nodes 0 to N as $[0 \rightleftharpoons N]$. Then, we can construct $[0 \rightleftharpoons N]$ from $[0 \rightleftharpoons (N-1)]$ via induction as follows.

$$[0 \rightleftharpoons N] = \qquad (8)$$
$$[0 \rightleftharpoons (N-1)] \to N \to 1 \to N \to 2 \cdots N \to (N-1) \to N \to 0$$

Let F_{N+1} denote TileIndex_{1D} with edges 0 to N. Essentially, $F_{N+1}(e_1, e_2)$ indicates the position of edge pair $e_1 \to e_2$ within $[0 \rightleftharpoons N]$. It can be derived by a similar induction process, as shown below.

$$F_{N+1}(e_1,e_2) = \begin{cases} F_N(e_1,e_2) & e_1 \leq N-1, e_2 \leq N-1 \\ {e_1}^2 + 2 \times e_2 - 1 & e_1 = N, 1 \leq e_2 \leq N-1 \\ 2 \times e_1 + {e_2}^2 & e_2 = N, 0 \leq e_1 \leq N-1 \\ (e_1+1)^2 - 2 & e_1 = e_2 = N \\ (e_1+1)^2 - 1 & e_1 = N, e_2 = 0 \end{cases} \qquad (9)$$

Finally, Equation 6 can be derived by unrolling Equation 9 through successively smaller values of N. You can also verify the correctness of Equation 6 by plugging it into Equation 9.

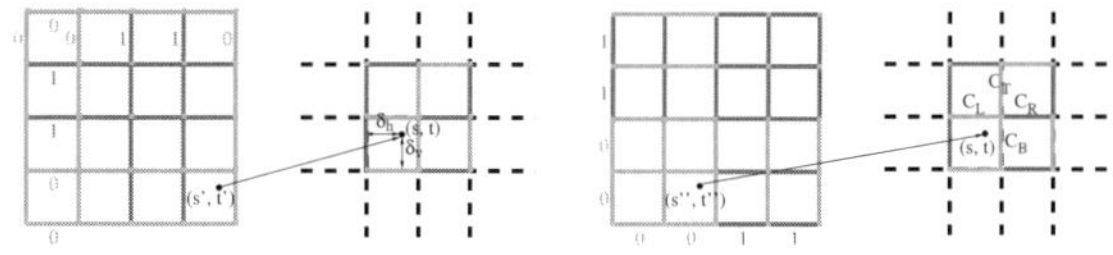

(a) Original Tile Packing (b) Dual Corner Packing

Figure 6: Dual packing examples.

Appendix B: Corner Handling

We derive the corner packing, CornerPack, by observing that CornerPack can be considered as the "dual" of TilePack. Since TilePack contains tiles with all possible edge color combinations via Equation 7, we can use the the same mechanism to pack corners with all possible edge color combinations. Let C_B, C_R, C_T, C_L be the bottom, right, top, and left edge colors around a corner. We can pack the corners as follows.

$$(J_h, J_v) = \text{CornerIndex}(C_B, C_R, C_T, C_L) = (\text{TileIndex}_{1D}(C_L, C_R), \text{TileIndex}_{1D}(C_B, C_T)) \qquad (10)$$

The computation is very similar to Equation 7. The major difference is that in Equation 7, the vertical/horizontal tile locations are determined by horizontal/vertical tile edge colors, while in Equation 10 the vertical/horizontal tile locations are determined by vertical/horizontal corner edge colors. As a corollary, the height/width of CornerPack is the same as the width/height of TilePack. CornerPack also inherits all the desirable properties from TilePack: (1) each corner appears only once in CornerPack, (2) adjacent corners in CornerPack have matching edge colors, and (3) the computation is as efficient as TilePack. An example of this dual corner packing is shown in Figure 6.

The computations involved in accessing CornerPack is also very similar to the access of TilePack shown above, with some minor differences due to the fact that we are locating the nearest corner rather than the containing tile. The steps are as follows.

- Given a texel request (s, t), we first compute which corner (Q_h, Q_v) is the nearest.

$$(Q_h, Q_v) = \text{floor}\Big((\text{s, t}) \times (M_h, M_v) - (\tfrac{1}{2}, \tfrac{1}{2})\Big) \,\%(M_h, M_v) \qquad (11)$$

Note the similarity of the computation of (Q_h, Q_v) with that of (O_h, O_v) in Equation 1. The only difference is the shift by $(\frac{1}{2}, \frac{1}{2})$ which enables floor() to locate the nearest corner for (s, t).

- Use (Q_h, Q_v) to hash the colors of the 4 edges at bottom, right, top, and left of the corner.

$$\begin{aligned} C_B &= \text{hash}[(Q_h+1)\,\%M_h + \text{hash}[2 \times Q_v]]\,\%K_v \\ C_R &= \text{hash}[\text{hash}[(Q_h+1)\,\%M_h] + (Q_v+1)\,\%M_v]\,\%K_h \\ C_T &= \text{hash}[(Q_h+1)\,\%M_h + \text{hash}[2 \times ((Q_v+1)\,\%M_v)]] \\ C_L &= \text{hash}[\text{hash}[Q_h] + (Q_v+1)\,\%M_v]\,\%K_h \end{aligned} \qquad (12)$$

This corner edge color computation is consistent with the tile edge computation in Equation 2 and is derived from it by proper changing of variables. For example, if (s, t) locates near the NE corner of a tile, then C_B computed in Equation 12 will be the same as C_E computed in Equation 2.

- For corner (C_B, C_R, C_T, C_L), use Equation 10 to locate its index (J_h, J_v) in CornerPack. Observe that (J_h, J_v) also indicates the location of the tile in CornerPack which has (C_B, C_R, C_T, C_L) as the NE corner.
- Derive the relative offset of (s, t) with tile (J_h, J_v).

$$(\varepsilon_h, \varepsilon_v) = \text{fraction}\Big((s, t) \times (M_h, M_v) - (\tfrac{1}{2}, \tfrac{1}{2})\Big) + (\tfrac{1}{2}, \tfrac{1}{2}) \quad (13)$$

Note the similarity of this computation with the computation of (δ_h, δ_v) in Equation 4. The only difference is the shift by $(\frac{1}{2}, \frac{1}{2})$ which enables us to compute the correct offset of (s, t) relative to the SW corner of tile (J_h, J_v).

- Fetch the desired texel from CornerPack.

$$\text{Tex}\left(\frac{J_h+\varepsilon_h}{P_v}, \frac{J_v+\varepsilon_v}{P_h}\right) \quad (14)$$

This value fetched from CornerPack is then interpolated with the texel fetched from TilePack to obtain the final texel value. The interpolation depends on how close is (s, t) to a tile corner.

Appendix C: The Cg Program of Our Algorithm

Some math symbols shown below are program constants as defined in Section 4. For simplicity, we have skipped the code for corner handling.

```
struct FragmentInput
{
  float4 tex : TEX0;
  float4 col : COL0;
};

struct FragmentOutput
{
  float4 col : COL;
};

float2 mod(const float2 a, const float2 b)
{
  return floor(frac(a/b)*b);
}

float EdgeOrdering(const float x, const float y)
{
  float result;
  if(x < y) result = (2*x + y*y);
  else if(x == y)
    if(x > 0) result = ((x+1)*(x+1) - 2);
    else result = 0;
  else
    if(y > 0) result = (x*x + 2*y - 1);
    else result = ((x+1)*(x+1) - 1);
  return result;
}

float2 TileLocation(const float4 e)
{
  float2 result;
  result.x = EdgeOrdering(e.w, e.y);
  result.y = EdgeOrdering(e.x, e.z);
  return result;
}

float4 Hash(uniform samplerRECT hashTexture,
    const float4 input)
{
  return texRECT(hashTexture,
      frac(input.xy/HASH_SIZE) * HASH_SIZE);
}

FragmentOutput fragment(const FragmentInput input,
    uniform sampler2D tilesTexture,
    uniform sampler2D cornersTexture,
    uniform samplerRECT hashTexture)
{
  FragmentOutput output;
  float2 mappingScale = float2(Mh, Mv);
  float2 mappingAddress = input.tex.xy * mappingScale;
  float4 numColors = float4(Kh, Kv, Kh, Kv);
  float2 thisVirtualTile = mod(mappingAddress, mappingScale);
  float2 nextVirtualTile = thisVirtualTile.xy + float2(1, 1);
  nextVirtualTile = frac(nextVirtualTile/mappingScale)*mappingScale;
  float4 edgeColors;
  edgeColors.x = Hash(hashTexture,
      Hash(hashTexture, thisVirtualTile.x) + thisVirtualTile.y);
  edgeColors.y = Hash(hashTexture,
      nextVirtualTile.x + Hash(hashTexture, 2*thisVirtualTile.y));
  edgeColors.z = Hash(hashTexture,
      Hash(hashTexture, thisVirtualTile.x) + nextVirtualTile.y);
  edgeColors.w = Hash(hashTexture,
      thisVirtualTile.x + Hash(hashTexture, 2*thisVirtualTile.y));
  edgeColors = frac(edgeColors/numColors)*numColors;
  float2 inputTile = TileLocation(edgeColors);
  float2 tileScale = float2(Ph, Pv);
  float2 tileScaledTex = input.tex.xy * float2(Mh/Ph, Mv/Pv);
  output.col = tex2D(tilesTexture,
      (inputTile.xy + frac(mappingAddress))/tileScale,
      ddx(tileScaledTex), ddy(tileScaledTex));
  return output;
}
```

Notice the use of ddx and ddy in the last tex2D() call. This is necessary to avoid spuriously large texture coordinate derivatives near tile boundaries which can cause low resolution mipmap levels to be accessed and produce over-blurry results.

Graphics Hardware (2004)
T. Akenine-Möller, M. McCool (Editors)

Silhouette Maps for Improved Texture Magnification

Pradeep Sen

Stanford University

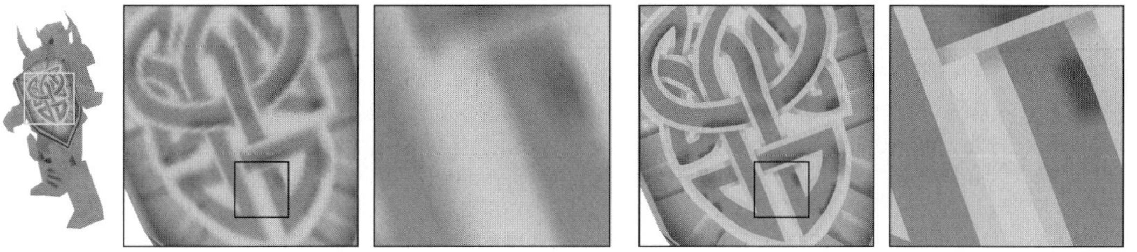

Abstract

Texture mapping is a simple way of increasing visual realism without adding geometrical complexity. Because it is a discrete process, it is important to properly filter samples when the sampling rate of the texture differs from that of the final image. This is particularly problematic when the texture is magnified or minified. While reasonable approaches exist to tackle the minified case, few options exist for improving the quality of magnified textures in real-time applications. Most simply bilinearly interpolate between samples, yielding exceedingly blurry textures. In this paper, we address the real-time magnification problem by extending the silhouette map algorithm to general texturing. In particular, we discuss the creation of these silmap textures as well as a simple filtering scheme that allows for viewing at all levels of magnification. The technique was implemented on current graphics hardware and our results show that we can achieve a level of visual quality comparable to that of a much larger texture.

Categories and Subject Descriptors (according to ACM CCS): I.3.1 [Computer Graphics]: Hardware Architecture I.3.3 [Computer Graphics]: Picture/Image Generation

1. Introduction

Texture mapping was developed by Catmull as simple a way of improving realism without increasing the geometric complexity of a scene [Cat74]. Since then, texturing has become a ubiquitous primitive in computer graphics, employed with much success in both high-end rendering for feature film as well as real-time graphics applications such as video games.

The sampling process inherent in computer graphics mandates that one must sample these textures carefully or risk introducing unsightly artifacts, a fact noted by Catmull in his original work. These can occur when the rendering samples in screen space coincide poorly with the discrete texture samples, eg. when the texture is magnified or minified on the screen. Although there has been extensive work for ameliorating artifacts during texture minification, relatively little work has been done in the regime where the texture is magnified, particularly for interactive applications. In this paper, we extend the silhouette maps introduced by Sen et al. [SCH03] to reduce the artifacts of textures under magnification. Our solution targets real-time applications and is amenable to acceleration on graphics hardware.

1.1. Previous work

We will begin with a survey of previous work specifically relating to the antialiasing of textures, and thus refer readers who seek an introduction to texture mapping to Heckbert's survey of the subject [Hec86]. As stated in the introduction, most of the previous work deals with textures under minification. One of the seminal papers in this area is Williams's work on mipmaps [Wil83]. A mipmap is a pyramidal data structure that maintains pre-filtered versions of a texture at different resolutions. The base layer contains the texture at full resolution, and each successive layer in the pyramid is $1/2$ the resolution of the layer below it. Texels in upper layers are computed by applying a simple box filter to the 2×2 texel neighborhood in the layer below it. To query for a filtered texture value, the algorithm must compute the size d of the pixel's projection onto the texture and then use this value to find the two neighboring mipmap levels with texels closest to that size. Texels from these two levels are then each bilinearly interpolated using the u and v texture coordinates, and the two results are linearly interpolated using d. The linear interpolation between levels prevents popping of the texture as the object moves back and forth. The main advantage of mipmaps is that they allow for a constant-time lookup of a filtered texture value regardless of the size of the pixel's footprint. They are also straightforward to implement in graphics hardware and are thus popular in real-time applications.

There have been other proposed approaches for accelerating the filtering of minimized textures for real-time applications (eg. Crow's summed-area tables [Cro84]). However, these approaches have not gained as widespread use as the mipmap algorithm in today's interactive applications.

So far, we have discussed previous work dealing with the regime where the texture is minified, or in mipmapping parlance the region of $d > 0$. Unfortunately, the problem of antialiasing of magnified textures has not received as much attention. The naïve approach is to simply perform a nearest lookup on the texture, resulting in a jagged "pixelated" texture. An improvement of this is to bilinearly interpolate the data, which results in blurry textures but alleviates pixelation considerably. Since this interpolation is fast and simple to implement (and supported natively on modern graphics hardware), most interactive applications use this technique despite low quality results. Finally, there is also the possibility of a brute-force approach which would simply store a higher resolution texture to begin with so that the user is never able zoom in beyond the base level in the mipmap. To reduce the memory consumption, compression techniques such as rendering from compressed textures [BAC96] could be implemented. Our proposed solution yields results comparable to that of a larger texture but with a modest memory overhead.

An approach that specifically addresses the magnification problem is SGI's GL_SGIS_sharpen_texture which tries to generate a higher resolution version of the texture by extrapolating from lower resolution versions. Unfortunately, without any more information other than the pixel values this extrapolation is often prone to error.

Schilling et al. describe detail maps in their Texram paper[SKS96], which provide a hierarchical resolution framework that allow certain regions of the texture to be stored at much higher resolutions than others. Unfortunately, because of the potentially large chain of indirect texture references, this method could be difficult to implement efficiently on consumer graphics hardware and can inherently only improve the quality of a few regions of the texture. Similar detail textures were implemented by SGI in their Infinite Reality system.

Another option is to use procedural textures [EMP*94]. Since these have a functional representation as opposed to a bitmap representation, they can be magnified arbitrarily. These textures tend to work well for some natural phenomena such as clouds, but are typically not well-suited for the textures normally found on signs, characters, etc. because it is often impossible to describe their artwork in functional form. They can also be computationally expensive, and all but the simplest are unsuitable for real-time.

This paper uses discontinuity information for improving filtering to reduce aliasing, something that is not new to graphics. One example of this was the successful use of discontinuity meshes for improved radiosity algorithms in the early 90's ([Hec92, LTG92]). On a more related topic, Salisbury et al [SALS96] described a piece-wise linear structure to allow magnification of pen-and-ink illustrations. Their technique is more complicated than the algorithm presented here and is not easily applied to current graphics hardware.

Bala et al. use an edge-and-point image, a representation that stores both samples and discontinuity information, to improve interpolation of sparse samples for high-end rendering [BWG03]. In work done concurrently to our own, Bala et al. extend their idea to texturing [RBW04]. Their technique, known as feature-based textures, approximates discontinuities in each cell with splines and can store several discontinuities in each cell. Feature-based textures target applications requiring high reconstruction quality, and it is not easy to see how the spline intersection and filtering parts of their algorithm could be mapped to graphics hardware.

Finally, in work done concurrently with our own, Tumblin et al. embed boundary information into a bitmap to create what they call bixels[TC04]. Bixels can be thought of as a silhouette map with two bits of extra information to indicate the boundary conditions. This is similar to our proposed extension of silhouette maps. The main difference in the two approaches lies in the reconstruction filter, with bixels using a more complex patch-based scheme. Our algorithm is simpler and has been implemented directly on the graphics hardware.

Our contribution is an algorithm that significantly improves the quality of magnified textures at comparable memory sizes and can be implemented on current graphics hardware. We do this by extending the silhouette map work of Sen et al. [SCH03] to general textures. While the standard bilinear interpolation attempts to reconstruct the original texture by applying a bilinear filter to all the bitmap samples, we improve the reconstruction by enhancing the point samples with a silhouette map that contains boundary information. We modify the original silhouette map algorithm to filter neighboring samples while preserving the discontinuities. Finally, we describe a simple way of antialiasing the texture when the textured object is minified. This results in a texture that can be viewed from up close or from far away without a degradation in quality and with a memory footprint com-

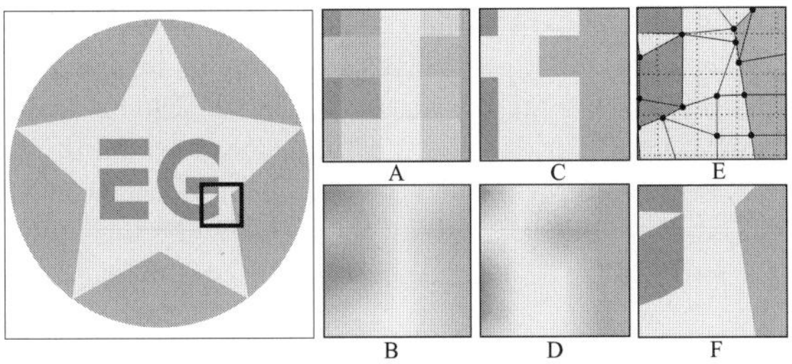

Figure 1: *This design was converted into a texture 32×32 pixels in size. The region indicated is a little over 3 pixels on a side. (A) Filtered bitmap (32×32) generated from the original vector representation. (B) Bilinear interpolation performed on the filtered bitmap during rendering. (C) Closeup of the bitmap used by the silmap algorithm, which is different than (A) because we do not pre-filter across discontinuities. (D) The same bitmap as (C), but this time bilinearly interpolated. (E) Positions of the silhouette points in the texels, with the deformed cells shown. Original silmap cells are in dashed lines. (F) Final result of our algorithm.*

parable to that of standard texturing. We call these silhouette map textures, or *silmap textures* for short.

2. Silhouette maps for textures

Silhouette maps were introduced by Sen et al. as a way to remove the jagged artifacts from the shadow mapping algorithm [SCH03]. Standard shadow maps store visibility from the light source in a depth texture which is then projected onto the scene during the final rendering pass. This projection often enlarges the depth texture in screen space, magnifying the depth texels and resulting in visible jagged contours along shadow boundaries.

The shadow silhouette map algorithm ameliorates these boundary artifacts by storing additional information about the position of the shadow edge in a data structure with fast, constant-time lookup. Specifically, a silhouette map is a uniform grid structure that stores the coordinates of single point in each cell. Every cell has a default point at the cell's center, (0.5, 0.5) in local coordinates, except when the silhouette boundary of an object passes through the cell. In this case, the point is placed on the silhouette boundary, and collectively the points form a piece-wise linear approximation to the shadow contour. This yields a better approximation of the true shadow contour than the piece-wise constant approximation achieved with standard shadow maps. To shade a point in the final rendering pass, it is first projected into the appropriate silhouette map texel. Lines connecting the cell's silhouette point to those of its four neighbors divide the cell into four *skewed quadrants*, and the point must be shaded according to the depth test result of its skewed quadrant.

At the end of the paper, the authors observe that a silhouette map can be thought to deform a regular grid to fit discontinuities. In this paper, we build on this idea and apply

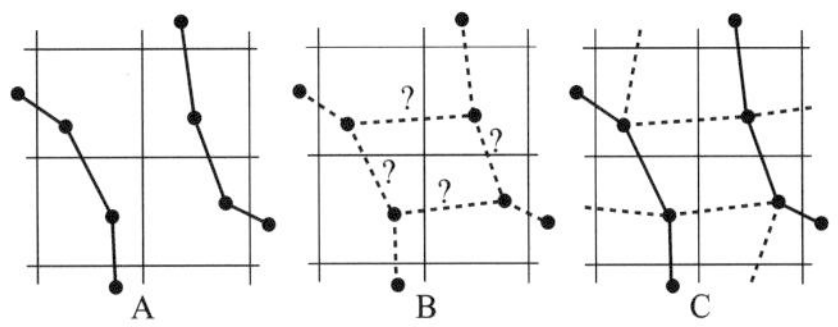

Figure 2: *(A) Situation in which two discontinuities are in close proximity to each other. (B) If only the discontinuity positions are stored in each cell, one cannot determine the true orientation of the boundaries. (C) Our solution is to store an extra bit for each edge to indicate whether the edge is valid. Here the valid edges are shown in a solid line.*

silhouette maps to general texturing. Without much modification, the silhouette map algorithm can perform nearest-neighbor lookups for textures, just like it did for the depth map. To do this, we think of the silhouette map as a more flexible structure than originally envisioned by Sen et al. Instead of storing scalar depth values in the cells of this deformed grid, we can easily store RGB color values (or any kind of values, for that matter). Thus, to render silmap textures we need two data structures at run-time: a regular bitmap, and a silhouette map that contains the edge information of the texture. As we will discuss later, these can be generated manually or automatically acquired from real photographs. The bitmap and the silhouette map are of the same resolution, and the bitmap is shifted 1/2 pixel in x,y with respect to the silhouette map so that the samples of the bitmap lie on the corners of the silhouette map cells.

During the rendering pass, the hardware projects the point to be shaded into the deformed bitmap and the deformed cell's sample is used as the final color for the sample. This yields results such as that shown in Fig. 1, where we can see that despite the low-resolution of the silmap texture we still get faithful reconstruction of the original texture. Since each deformed cell can only have one color, this algorithm is better suited for textures with regions of constant colors, such as decals and signs than for general textures. Besides RGB, one can also store alpha values for transparency or other variables so that shaders can be specified with high resolution using a relatively low-resolution data structure. Details on the implementation of this algorithm can be found in a later section.

2.1. Filtering

The nearest-neighbor algorithm is satisfactory for only a small number of applications. The problem with it is evident when we apply the algorithm to more complex textures (Fig. 5). If we do not filter neighboring samples together, we see that the interior regions appear pixelated. To reduce these artifacts, we must filter across a neighborhood of samples while being careful not to filter across established discontinuities. We must also keep the samples localized enough so that they can be accessed quickly on graphics hardware.

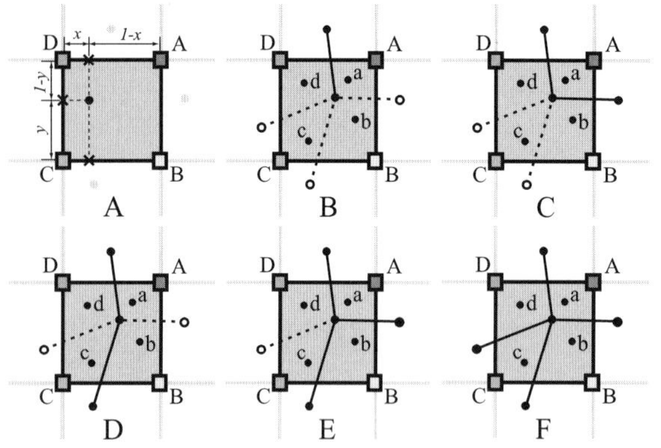

Figure 3: *Cases requiring different interpolation schemes. Corner samples representing bitmap entries are labeled A, B, C, D and random points are shown in each skewed quadrant to facilitate discussion. (A) Cell without boundaries, so the color is determined by bilinearly interpolating the four samples. (B) Cell contains one boundary segment. (C,D) Cell contains two boundary segments. (E) Three boundary segments, and finally (F) all four border segments.*

In order to avoid filtering across discontinuity boundaries, we must be able to define these boundaries in our silhouette map. The problem with the silmaps of Sen et al. is that they contain the position of potential discontinuities but do not indicate what is a real boundary [SCH03]. In their implementation, the depth tests at the corners are used to establish boundary information since shadow boundaries can only occur along an edge where the two depth tests give conflicting results. In our situation we do not have these tests so we must add additional information to the silhouette map structure to indicate these boundaries. One might be tempted to address this problem by only having silhouette points inside cells with valid discontinuities and empty cells elsewhere, with the hope of extrapolating boundary information by looking at neighboring cells. This approach does not work, however, since it can yield ambiguous situations (Fig. 2). The fundamental problem is that while the position of the discontinuity can be interpreted as a property of the cell, the connectivity of the discontinuity boundary of the cell with its neighbors is a property of the edges. Therefore, there can be up to 4 boundary edges in each cell.

Our solution is to add two extra bits per cell, which are then associated with two of the four boundary segments and indicate whether they represent a discontinuity or not. Only two of the four segments need to be labeled per cell because we can access the information on the other two from neighboring cells. This method specifies the boundaries of the texture explicitly and breaks the texture up into *regions*, sets of points located on the same side of the discontinuity that must be filtered together. In this paper, we will refer to the boundaries of a cell as the north, south, east, and west boundaries depending on the neighbor they connect to.

Initially we experimented with storing a region code at each corner of the cell to indicate its region instead of explicitly defining the boundaries. As in the shadow application, the boundaries could then be deduced from this information since corners with different region codes would have an implicit boundary in between them. While this allowed the re-use of the nearest neighbor lookup code to fetch the region information in the same way we fetched the color information in the previous section, problems were encountered when a region curved around and had a boundary with itself. This required arbitrarily breaking up a continuous region into two different regions in order to preserve the boundary. Because of this, we found it better to explicitly express the boundaries in each cell.

When rendering with a silmap texture, we must also determine how many samples from the bitmap to use in texture reconstruction. More complicated reconstruction schemes use samples from neighboring cells [RBW04, TC04], but this makes them more difficult to implement on graphics hardware. In this work, we decided to keep the samples localized and will work only with the four corner color samples of a silhouette cell. Not only is this realizable in current hardware, but it is fast and provides reasonable results. For the same reason, we chose bilinear interpolation as our filter basis. It works well for real-time applications and is supported natively in the hardware, a fact that we used to our advantage in the implementation.

At this point we can describe the full silhouette map texture algorithm with filtering. First, the *uv* coordinates of the point to be textured are computed to project it into a specific cell of the silmap. We must then decide which of the four corner samples are in the same region as the current sample by using our boundary information (Fig. 3). Once we have decided which corners to include in the region, we must apply the appropriate filter to the samples to get the final color value.

First, we determine which corner samples are available for filtering by following these rules:

1. A point is always in the same region as its respective corner. For example, in Figure 3, point a will always include corner sample A in its filter kernel.
2. Corner samples sharing a side with the point's corner will be included only if there is not a boundary in the way. For example, point a will use corner sample B only if there is no *east* boundary. Likewise it will include corner sample D if there is no *north* boundary.
3. The corner sample directly across the point's corner will be included only if there is a path there that does not intersect boundaries. For the case of point a we can write the decision to include corner sample C as the following boolean expression:

```
inc_C = ((!b_N && !b_W) || (!b_E && !b_S))
```

where boolean b_N is a 1 if there is a *north* boundary, for example. Another way to say this is that corner sample C

will be included if we can move counter-clockwise from A without hitting a boundary (*north* and *west* boundaries are both 0) *or* we can move clockwise without hitting a boundary (*east* and *south* boundaries are both 0).

To better understand this algorithm, we examine specific examples from Figure 3. In Fig. 3B, we can see that point d will include sample D by the first rule. In addition, it will include C and dismiss A by rule 2. Finally, since we can access B from D (because the *west* and *south* boundaries are both 0), B will be included as well. Thus, point d will filter between three corner samples, B, C, and D. In the same diagram, it can be seen that point c will include all four corner samples by following the rules stated. This means that in this example we could end up with a shading discontinuity along the non-existent *west* boundary because different filters are used on either side of the "boundary." While this is an artifact of the simple filtering algorithm, we note that in practice the results are reasonable and schemes that guarantee not to have these artifacts for any case are difficult to implement for real-time.

At this point we have determined how many corners (from one to all four) we will use in the interpolation. Because we are using a bilinear basis for filtering, several of the cases become straightforward to implement on graphics hardware. For the 1-corner case (eg. point b in Fig. 3E), we simply use the value of the appropriate corner. For the 2-corner case (eg. point c in Fig. 3D), we linearly interpolate between the two corner values, and with the 4-corner case we perform the full bilinear interpolation.

The 3-corner case requires special consideration, however. Suppose we wanted to find the value for point c in Fig. 3C using samples B, C, and D. We first compute the plane going through these samples by performing the cross products of the two vectors

$$\vec{CB} = (1,0,C_B - C_C), \vec{CD} = (0,1,C_D - C_C)$$

where C_B indicates the color at corner B. The cross product of them yields the normal

$$\begin{aligned} \vec{N} = \vec{CB} \times \vec{CD} &= \begin{vmatrix} \vec{i} & \vec{j} & \vec{k} \\ 1 & 0 & C_B - C_C \\ 0 & 1 & C_D - C_C \end{vmatrix} \\ &= -(C_B - C_C)\vec{i} - (C_D - C_C)\vec{j} + \vec{k} \end{aligned}$$

and so we have a plane equation which is equal to:

$$\vec{N} \begin{bmatrix} x - x_0 \\ y - y_0 \\ z - z_0 \end{bmatrix} = \vec{N} \begin{bmatrix} x \\ y \\ z - C_C \end{bmatrix} = 0$$

and which simplifies to

$$z = (1 - x - y)C_C + xC_B + yC_D$$

This is also the same result as barycentric coordinates for our triangle in the unit square.

We can now write down the rules for filtering each of the different cases in Fig. 3. Without loss of generality we write down the formulas for the cases specifically as shown. In this paper, we refer to (x,y) to be the local coordinates with respect to the origin of the texel, corner C. The global texture coordinates are (u,v).

Case	Equation
1 corner	C_C
2 corner	$(1-x)C_D + xD$
3 corner	$(1-x-y)C_C + xC_B + yC_D$
4 corner	$(1-y)((1-x)C_C + xC_B) + y((1-x)C_D + xC_A)$

Now we must come up with a way so that all of the different configurations can be handled efficiently in the hardware. In the implementation section, we discuss how we take advantage of the bilinear interpolator available on the hardware to perform this filtering in an efficient manner.

2.2. Mipmapping of silhouette map textures

Since silhouette maps introduce higher frequency components to the original texture, it is important to filter them properly when they are minimized or they will alias when the textured surface is animated.

Our solution to this problem takes advantage of the fast mipmapping capabilities of the current graphics hardware. After the silhouette map and the respective bitmap are created, a pre-processing step determines an average color for every cell in the silhouette map by weighting the corner samples by the area coverage of each of their respective quadrants. This generates a bitmap with filtered colors at the same resolution as the original silhouette map. This texture can then be mipmapped like a standard bitmap. The mipmap is generated from the silmap texture itself instead of simply using the original bitmap to ensure that the two are perfectly aligned.

Because the frequency content introduced by the silmap texture can be quite high, we switched over in some of our examples to the mipmapped version as soon as the screen/texture sample ratio became 1:1 to avoid aliasing. In the implementation section we describe how we use the mipmapping hardware to switch between the silmap and mipmapped textures.

2.3. Authoring of silhouette map textures

Silhouette map textures can be created in many ways. A drawing tool that combines bitmap and vector graphics would allow artists to manually create the silmap directly. Silmap textures can be generated from photographs by either extracting the discontinuity edges through standard computer vision techniques (eg. [Can86]) or manually annotating the edges. Once the edges are annotated, the original image can be filtered down to the appropriate size for the silhouette map while respecting the discontinuity edges. It is essential that one does not filter across discontinuities when generating the bitmap data for the silmap texture (Fig. 6). In a way, the process of generating a silmap texture from an original image is similar to the texturing process.

2.4. Memory usage

The full silhouette map texturing algorithm proposed in this paper requires three pieces of information per texel: the color information from the bitmap, a silhouette point that encodes the position of the discontinuities, and the explicit identification of boundaries.

In order to reduce the memory overhead, the silhouette map and the region map can be packed into a single byte by using three bits for both x and y. This gives us 8 quantized locations in x and y, for a total of 64 possible locations inside the texel. The remaining two bits can be used for directly labeling the edges for discontinuities.

3. Implementation

3.1. Nearest-neighbor silhouette map textures

The nearest-neighbor silhouette map texture algorithm was implemented in ARB_fragment_program code. The algorithm is shown below (including instruction counts):

1. Fetch current silhouette point and 4 neighbors. [5 TEXs]
2. Translate neighboring points into the cell's local coordinate frame. [4 ADDs]
3. Fetch corner color samples. [4 TEXs]
4. Compute line equations for the 4 boundary segments and for 3 of the 4 line segments that connect the point to the corners of the cell. These corner line segments divide 3 quadrants in the cell into 2 slices each. The last quadrant is not tested since it is the default. [7 XPDs]
5. Test the point to shade against each line to see where it lies. [7 DP3s]
6. Move the correct color into the output depending on the quadrant we are in. [2 CMPs/slice × 2 slices/quadrant × 3 quadrants (last one is default)]

The explanation as to why we also test against the corner line segments is given in Figure 4.

3.2. Filtered silhouette map textures

For the hardware implementation of the filtered silmaps, we must first determine which corners are in the same region as the sample point. Since we do not have conditional execution, we must assume that our point can be any of the generic points a, b, c, or d in Fig. 3. A *reachability* vector (to use Bala's terminology [BWG03]) V is computed whose components indicate which corner samples should be included in the order $\{A,B,C,D\}$. This vector is generated for each of the generic points, and then the nearest neighbor algorithm is used to determine which skewed quadrant the point is in

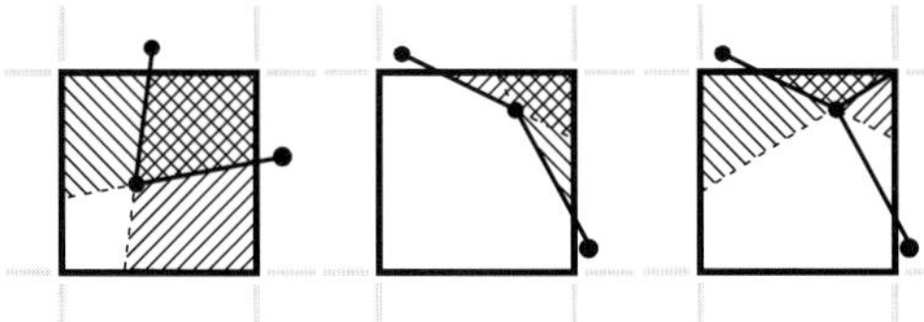

Figure 4: *Taking the dot product of the point to shade with the boundary lines to identify the quadrant only works when the angle formed by these lines is less than 180°. In this diagram, the hatched patterns indicate the regions where the dot product is of the same sign for each of the two boundary lines. (A) When the angle is less than 180°, the entire skewed quadrant is covered by the two tests, therefore we can simply AND the result of the two tests to determine if the point is in this quadrant. (B) If the angle becomes greater than 180°, however, the cross-hatch does not cover the entire quadrant and our test would give false negatives in regions with only one hatched pattern. To avoid these problems, we also test against the lines to the corners of the cell, turning each quadrant into two pie-shaped slices. (C) Here we show the results of testing against the first slice in the NE quadrant. This slice is guaranteed to be covered by the two tests.*

in order to move the correct V into the proper variable to be used in the filtering portion of the algorithm. As an example, the reachability vector for point a would be:

```
Va.x = 1 //A is always reachable
Va.y = !b_E // include B if no east boundary
// the C corner is more complicated, as we discussed...
Va.z = (!b_N && !b_W) || (!b_E && !b_S)
Va.w = !b_N // include D if no north boundary
```

Similar vectors can be written for the other quadrants. We must now implement the filtering equations $f(x,y)$ for each of the cases in an efficient manner for all the possible reachability vectors. We observe that it is possible to accelerate the computation in every one of these cases by using the bilinear interpolation already available in the hardware:

$$g(x,y) = (1-y)((1-x)C_C + xC_B) + y((1-x)C_D + xC_A)$$

In the 1-corner case we simply want to use the color sample at the corner. Suppose that this is corner C. We can access this value with $g(0,0)$. If we have a 2-corner situation, say corners B and C, we can access the result of the linear interpolation through $g(x,0)$. Finally, the 4-corner computation is the same as simply accessing $g(x,y)$ directly.

The 3-corner case requires more work. Take as an example point c in Fig. 3C. We have determined that for this case, $f(x,y) = (1-x-y)C_C + xC_B + yC_D$. We note that:

$$g(x,0) = (1-x)C_C + xC_B$$

$$g(0,y) = (1-y)C_C + yC_D$$

Configuration				
A	B	C	D	f(x,y)
0	0	0	0	invalid
0	0	0	1	g(0,1)
0	0	1	0	g(0,0)
0	0	1	1	g(0,y)
0	1	0	0	g(1,0)
0	1	0	1	invalid
0	1	1	0	g(x,0)
0	1	1	1	g(x,0) + g(0,y) - g(0,0)
1	0	0	0	g(1,1)
1	0	0	1	g(x,1)
1	0	1	0	invalid
1	0	1	1	g(x,1) + g(0,y) - g(0,1)
1	1	0	0	g(1,y)
1	1	0	1	g(x,1) + g(1,y) - g(1,1)
1	1	1	0	g(x,0) + g(1,y) - g(1,0)
1	1	1	1	g(x,y)

Table 1: *Computing the interpolated value using only g(x,y)*

which when added give us

$$\begin{aligned} g(x,0)+g(0,y) &= (1-x)C_C+xC_B+(1-y)C_C+yC_D \\ &= 2C_C-xC_C-yC_C+xC_B+yC_D \end{aligned}$$

so to get f(x,y) for this configuration, we need to simply subtract out the extra C_C term:

$$\begin{aligned} f(x,y) &= g(x,0)+g(0,y)-g(0,0) \\ &= C_C-xC_C-yC_C+xC_B+yC_D \\ &= (1-x-y)C_C+xC_B+yC_D. \end{aligned}$$

Thus it is possible to preform the 3-corner interpolation by performing a linear interpolation of the two sides and subtracting out the sample at the right-angle corner. This allows us to derive Table 1 which shows how to compute $f(x,y)$ using only $g(x,y)$ for all possible configurations. The 1's and 0's in the table indicate which corners are in the same region as the current sample. There are three invalid entries in this table. The corner the sample is in is automatically included, so we can ignore the all-zero case. The other two invalid configurations cannot occur in the silhouette map.

The table shows that the interpolated value can be accessed with one texture lookup for the 1, 2, and 4-corner cases and with 3 texture lookups for the 3-corner case. In the 3-corner case, we must be careful since it is possible that $f(x,y)$ could be bigger than any of the three colors involved (this can happen when the sample is outside the triangle formed by the three corners). In this case we perform a per-component clamp to the maximum component of the two corners on the diagonal of the triangle. For example, if we are dealing with the 3-corner situation we have looked at before, we would clamp to $max(C_B,C_D)$, where the *max* is done per-component. We found in initial experimentation that if we did not clamp in this manner, we would have problems in the cases where there was a large gradient between two of the corner samples of the triangle. This would yield a color that was noticeably incorrect yet still in the still in the range 0, 1.

To implement this in an ARB_fragment_program, we constructed the Karnaugh map of the function to compute the coordinates of each $g(x,y)$ term. For example, the x coordinate of the first $g(x,y)$ term is sometimes 0, sometimes 1 and sometimes x. By simplifying the Boolean logic that take the inputs *ABCD* and maps them to one of these values, we were able to come up with expressions that mapped well to the conditional move statements available in the hardware. This yielded a fragment program that was 45 instructions long.

3.3. Mipmapping

The mipmapped versions of the silhouette maps were computed as a pre-process. During rendering, certain portions of the texture might need to be mipmapped while others are being magnified at the same time. In order to switch between the silhouette map texture and the mipmapped version, we first rendered the mipmap levels into a texture in solid colors. This is similar to the way mipmap levels are visualized, but in our case only the base level was of a different color (Fig. 7). To prevent popping as the program switched between the silmap texture and the mipmapped version, we blended between levels to get a smooth transition. Our tests suggested that even without the smooth blending between levels the popping was not noticeable. Once we had established which regions of the screen we wanted to use with which algorithm, we used the this texture to combine the results of a silmap texture pass and a mipmap pass. The final result was a scene that would smoothly switch between the mipmap and the silmap texture as needed.

4. Results

Using a silhouette map editor we developed, we created several silmap textures and scenes to test our algorithm. Our implementation runs on both NVIDIA and ATI hardware at real-time rates. To give some concrete numbers, the knight scene with the full algorithm including blending and mipmapping runs at 70 fps on a pre-release ATI X800XT (500/500MHz). The knight is an animated md2 character and is textured by a silmap 256×256 texels in size. Images from this and other tests are shown throughout the paper. In places where we compare to standard bilinear interpolation, we use a prefiltered version of the bitmap (not the bitmap we incorporate into the silhouette map), since this is what you would normally use in practice and it makes the quantization artifacts less visible. This causes the images to look different even in regions where full bilinear interpolation occurs in both. When we substitute the bitmap from the silhouette map into the standard algorithm, the fully-blended regions are identical as one would expect.

Texture	Resolution	Original size	Mipmapped size	Silmap texture algorithm
Knight	256×256	196.6KB	262.1KB	524.3KB
Teddy Bear	64×64	12.3KB	16.4KB	20.5KB
EG logo	32×32	3.1KB	4.1K	8.2K
Pedestrian X-ing	128×128	49.1KB	65.5K	131.1K

Table 2: *Survey of memory usage for some of the textures we tested. Our algorithm yielded results equivalent to using a standard texture many times larger in size while typically only doubling the amount of memory required.*

Since our algorithm can be considered a form of texture compression, it is of particular interest to study the memory consumption of our new technique. A comparison with standard texture maps is shown in Table 2 for a few textures.

Our results show that our algorithm roughly doubles the size of the memory needed while at the same time providing an image quality of a texture many times larger. Increasing the size of the texture to improve quality is expensive since you can only increase the texture in $\sqrt{2}$ times in each direction before it doubles in size.

Note that the memory consumption of our silmap technique is based on theoretical calculations. In practice we had to use a larger texture because we cannot pack everything into a single byte (fragment programs do not have bitwise operations). If the algorithm had native hardware support then these numbers could be achieved.

5. Discussion

5.1. Artifacts

The artifacts associated with silhouette maps have been discussed by Sen et al [SCH03]. These artifacts occur when the discontinuity has a region of high curvature that cannot be represented by a single point or when multiple discontinuity curves intersect inside a cell. These artifacts can also occur in silmap textures. In addition, the simple interpolation scheme we propose can sometimes cause banding artifacts, especially when two cells with a different number of valid corners are next to each other. Our experimentation demonstrated that these artifacts are not objectionable.

Despite these constraints, it should be emphasized that silmap textures are created as a pre-process and, unlike the silhouette map shadows, they can be modified and adjusted to eliminate all artifacts. Once the silmap texture is artifact-free it is guaranteed to be correct at run-time. When creating the examples of this paper, we found that after some practice it became straightforward to avoid problematic situations.

5.2. Real photographs

The algorithm was tested on real photographs like those in Figures 5 and 6. While the results look reasonable, the sharp discontinuity boundaries tend to give the photographs a cartoon-like quality, especially when viewed under large magnification. The explanation of this phenomenon is not complicated. Natural photographs are composed of wide range of different frequencies. Our algorithm explicitly preserves discontinuities, which are stored as high frequency components in the frequency domain. Because we aggressively downsample the image, we filter away all of the middle frequencies and end up with a representation that preserves the low and high frequencies of the original photograph but lacks the mid-range. Our minds automatically interpret these images as being cartoon-like, since cartoons often consist of low-frequency regions of fairly constant color carefully separated from each other through high-frequency outlines. Thus this is not something particular to our algorithm, but inherent to all the algorithms that attempt to explicitly store high-frequency discontinuity information while filtering down the color information (eg. bixels and feature-based textures). This suggests that simply avoiding filtering across discontinuity boundaries is not enough for high quality rendering of enlarged photographs.

5.3. Other applications

The silhouette map textures we propose in this paper can be used for more than just simple texturing. In Figure 9, we use a texture to modulate a procedural noise shader [EMP*94] on a surface. Because the silmap texture gives us a piecewise linear approximation of the discontinuities of the design pattern, even a low resolution texture can give us precise control of the shader on the surface of the object. Another potential application of silmap textures are for billboarding, where the alpha channel is typically a boolean indicating transparency. Using current techniques, billboarded trees appear jagged when viewed from up close. Silmap textures would greatly reduce this problem.

5.4. Hardware acceleration

The silhouette map texture algorithm could benefit tremendously from hardware acceleration, even if only part of it is performed by the hardware. One can imagine a texture unit that automatically performs a nearest-neighbor lookup on a texture that has been enhanced with a silhouette map. Such a unit could be used to automatically generate the images in Figs. 1 and 8 without the need of a fragment program. It could also be accessed by the shader in Fig. 9 as easily as one does a texture fetch. To filter more complex textures, one can write a fragment program that relies on this unit to help compute the reachability vector to decide which corners to include in the filtering operation. Finally, this unit would allow the silhouette map shadow algorithm to access the depth map through the same mechanism, thus eliminating the need for a fragment program in the final pass altogether.

6. Further work and conclusion

Any future work on silhouette map textures should probably begin by exploring the application of the technique to natural photographs. As presented in this paper, our algorithm stores RGB values in each cell which we then interpolate together to get the final image. While RGB is one useful basis, other bases (eg. DCT coefficients or indices to texture patches) might be better suited for reproducing photographs. This could open the door to interesting hybrid techniques for image representation. On the more practical side, there needs to be hardware and software support for silhouette maps if the technique is to be adopted by the game development community. Therefore, a good editor that merges seamlessly into current production pipelines is a must.

This paper presented a simple algorithm that considerably improves the quality of textures under magnification while modestly increasing the memory required. The silmap textures are fairly easy to create and the algorithm runs on current graphics hardware.

7. Acknowledgments

The initial ideas that led to this work were conceived through discussions with Mike Cammarano and Pat Hanrahan while working on the paper on silhouette map shadows. Yasamin Mostofi, Mike Houston, and Tim Foley proofread the paper and gave valuable feedback during writing. Special thanks to Anup Lobo who beta-tested the silmap editor and generated the silmap for Figure 9. Thanks also to Fluff for willing to be photographed. The model for the knight shown throughout the paper is available at www.polycount.com and the engine to render it was based on the NVIDIA shadow volumes demo. We would also like to thank NVIDIA and ATI for providing the hardware and support needed to run our experiments. Pradeep Sen was supported by a research grant from ATI, NVIDIA, and SONY.

References

[BAC96] BEERS A. C., AGRAWALA M., CHADDHA N.: Rendering from compressed textures. In *Proceedings of the 23rd annual conference on Computer graphics and interactive techniques* (1996), ACM Press, pp. 373–378.

[BWG03] BALA K., WALTER B., GREENBERG D. P.: Combining edges and points for interactive high-quality rendering. *ACM Trans. Graph. 22*, 3 (2003), 631–640.

[Can86] CANNY F. J.: A computational approach to edge detection. *IEEE Trans PAMI 8*, 6 (1986), 679–698.

[Cat74] CATMULL E. E.: *A subdivision algorithm for computer display of curved surfaces.* PhD thesis, University of Utah, 1974.

[Cro84] CROW F. C.: Summed-area tables for texture mapping. In *Proceedings of the 11th annual conference on Computer graphics and interactive techniques* (1984), ACM Press, pp. 207–212.

[EMP*94] EBERT D., MUSGRAVE K., PEACHEY D., PERLIN K., WORLEY: *Texturing and Modeling: A Procedural Approach.* Academic Press, Oct. 1994.

[Hec86] HECKBERT P. S.: Survey of texture mapping. *IEEE Comput. Graph. Appl. 6*, 11 (1986), 56–67.

[Hec92] HECKBERT P. S.: Discontinuity meshing for radiosity. In *Rendering Techniques '92* (1992), Chalmers D. P. A., Sillion F., (Eds.), Eurographics, Consolidation Express Bristol, pp. 203–216.

[LTG92] LISCHINSKI D., TAMPIERI F., GREENBERG D. P.: Discontinuity meshing for accurate radiosity. *IEEE Computer Graphics and Applications 12(6)* (Nov. 1992), 25–39.

[RBW04] RAMANARAYANAN G., BALA K., WALTER B.: Feature-based textures. In *Proceedings of the Eurographics Symposium on Rendering* (2004), Eurographics Association.

[SALS96] SALISBURY M., ANDERSON C., LISCHINSKI D., SALESIN D. H.: Scale-dependent reproduction of pen-and-ink illustrations. In *Proceedings of the 23rd annual conference on Computer graphics and interactive techniques* (1996), ACM Press, pp. 461–468.

[SCH03] SEN P., CAMMARANO M., HANRAHAN P.: Shadow silhouette maps. *ACM Transactions on Graphics (Proceedings of ACM SIGGRAPH 2003) 22*, 3 (July 2003), 521–526.

[SKS96] SCHILLING A., KNITTEL G., STRASSER W.: Texram: A smart memory for texturing. *IEEE Comput. Graph. Appl. 16*, 3 (1996), 32–41.

[TC04] TUMBLIN J., CHOUDHURY P.: Bixels: Picture samples with sharp embedded boundaries. In *Proceedings of the Eurographics Symposium on Rendering* (2004), Eurographics Association.

[Wil83] WILLIAMS L.: Pyramidal parametrics. *SIGGRAPH Comput. Graph. 17*, 3 (1983), 1–11.

Graphics Hardware (2004)
T. Akenine-Möller, M. McCool (Editors)

PixelView: A View-Independent Graphics Rendering Architecture

J. Stewart, E.P. Bennett, and L. McMillan

Department of Computer Science, The University of North Carolina at Chapel Hill, USA

Abstract

We present a new computer graphics rendering architecture that allows "all possible views" to be extracted from a single traversal of a scene description. It supports a wide range of rendering primitives, including polygonal meshes, higher-order surface primitives (e.g. spheres, cylinders, and parametric patches), point-based models, and image-based representations. To demonstrate our concept, we have implemented a hardware prototype that includes a 4D, z-buffered frame-buffer supporting dynamic view selection at the time of raster scan-out. As a result, our implementation supports extremely low display-update latency. The PixelView architecture also supports rendering of the same scene for multiple eyes, which provides immediate benefits for stereo viewing methods like those used in today's virtual environments, particularly when there are multiple participants. In the future, view-independent graphics rendering hardware will also be essential to support the multitude of viewpoints required for real-time autostereoscopic and holographic display devices.

Categories and Subject Descriptors (according to ACM CCS): I.3.3 [Computer Graphics]: Viewing Algorithms; Display Algorithms; Bitmap and Frame Buffer Operations. I.3.1 [Computer Graphics]: Graphics Processors. I.3.6 [Computer Graphics]: Graphics Data Structures and Data Types.

1. Introduction

Viewpoint specification is fundamental to traditional computer graphics rendering. Both the transformation of a scene to eye space in the traditional graphics pipeline and the origination of viewing rays in a ray-casting system depend on the viewpoint. Moreover, many subsequent rendering steps are also impacted by the choice of viewpoint, including clipping, projection, illumination calculations, shading, and visibility determination. As a result, changing the viewpoint frequently gates the entire process of interactive rendering, as each rendered frame is initiated with the specification of a viewpoint, followed by the scene description, and culminating with the final displayed image.

There are many potential advantages to decoupling viewpoint specification from rendering. First, immediate efficiency improvements are available if rendering costs are amortized over multiple views. They result from reuse of shading calculations as well as exploiting the coherency of surface reflection with smooth variations in viewpoint. A second advantage results from beginning the rendering process before the viewing position is resolved, thereby reducing latency.

However, the ultimate advantage of separating rendering from viewpoint selection is that it becomes possible to render the same scene for multiple eyes. Possible applications include shared virtual environments (stereo viewing by many participants of a computer-generated scene). In the future, view-independent graphics rendering hardware will also be essential to support the multitude of viewpoints required for real-time autostereoscopic and holographic display devices.

Figure 1: *Photograph of our hardware prototype of the PixelView architecture. The prototype is on the right, its VGA output is in the middle, and the PC controller is on the left.*

We have developed a new computer graphics rendering architecture, called PixelView, in which a single traversal of the scene description generates "all possible views" (or at least a wide range of views). PixelView is compatible with the scene descriptions used by traditional 3D rendering hardware (i.e. polygons). In addition, PixelView, like the Reyes rendering system and many ray-casting renderers, supports any higher-order surface primitive for which a world-space subdivision scheme exists. It also natively supports point-based models as well as image-based representations, all within a unified rendering architecture.

The primary contributions of our PixelView rendering system include:

- A scalable system architecture for supporting real-time, view-independent rendering of 3-D models
- A hardware prototype of a 4D, z-buffered frame buffer demonstrating the viability of dynamic view selection at scan-out
- Hardware rendering of primitives with view-dependent reflectance by a technique that we call "point casting"

In addition, we consider possible representations and compressions for the radiance of view-dependent points to reduce storage and bandwidth requirements.

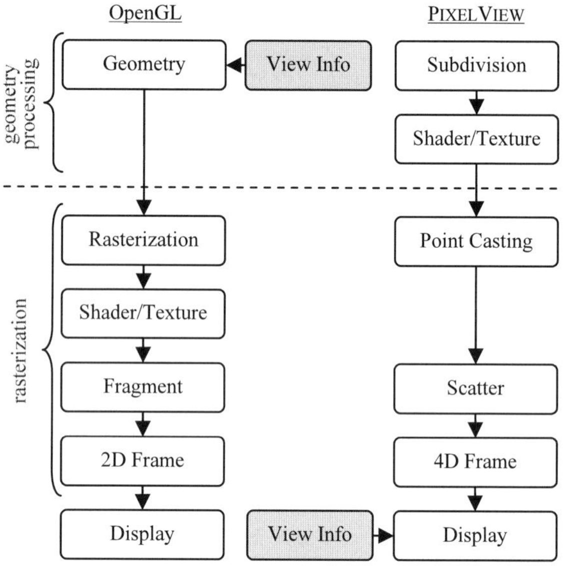

Figure 2: *High-level comparison between a typical OpenGL-style pipeline and PixelView. Only the display stage in PixelView requires knowledge of the viewpoint, allowing reuse of shading and rendering calculations.*

2. Overview

We begin with a high-level overview of PixelView via a comparison with the traditional graphics pipeline, with a focus on the role of view specification.

The left side of Figure 2 shows a traditional graphics pipeline. Polygon vertices (in object space) are fed into the geometry stage for transform and lighting. This stage outputs lit, screen-space triangles that are subsequently rasterized into fragments. These fragments are then shaded and textured. Various other raster operations, such as depth-comparison, compositing, and filtering, can then be performed before the final color values are written into the 2D frame buffer.

The right side of Figure 2 shows the PixelView pipeline. Our system supports any primitive that can be subdivided in world space. Primitives enter the subdivision stage, where they are first transformed from object space into world space. The primitives then undergo world-space subdivision and are shaded and textured. The shader/texture stage outputs fully shaded world-space points, which are then "point cast" into specific bins of the 4D frame buffer. Lastly, because a point will generally not fall exactly into a bin, its contribution is "scattered" into nearby bins. Specifics of the various pipeline stages are discussed in subsequent sections. For now, we will focus on the role that view information plays in each rendering approach.

When comparing the pipelines, note that view specification comes into play very early in the traditional graphics pipeline. Specifically, the first operation that typically happens to vertices in the geometry stage is to transform them into eye space via the modelview matrix. Many later steps in the pipeline are also affected by the viewpoint. Thus, changing the viewpoint requires re-traversal and re-rendering of the entire scene.

Alternatively, note that no view transform occurs in the PixelView geometry processing stages. In fact, no view transform occurs at any point along a primitive's path into the frame buffer. That is, no knowledge of the virtual camera's location and orientation is required to render primitives into the 4D frame buffer. We can "get away with this" because we render a sampled version of the outgoing radiance from each point of the scene. The viewpoint specification can thus move to the display stage, where it is used to reconstruct a particular view during scan-out. This allows the viewpoint to change without re-rendering the scene. Furthermore, this fundamental reorganization of rendering tasks allows us to re-visit the tradeoffs between rendering quality and frequency of scene updates.

3. Previous Work

The recent advent of flexible programmable graphics hardware has ignited renewed interest in alternative architectures for real-time rendering [OKTD02; PBMH02]. However, one of the least flexible remaining functions in existing architectures is the scan-out mechanism used to refresh the display. The only exposed controls for display refresh are for setting the resolution (and perhaps setting the origin of the memory block allocated for the frame buffer). There are compelling reasons for optimizing this functionality via a hardwired implementation. In particular, because the performance of graphics processing units is often dictated by available memory bandwidth, there is a significant benefit in optimizing the appreciable, constant, and real-time demands of display refresh. In other words, it makes sense to minimize the demands of refresh to free up memory bandwidth for other compelling rendering functions.

However, the lack of flexibility in display scan-out limits the ability to address certain problems related to dynamic view selection and latency compensation [OCMB95; BFMZ94]. Regan et al. recognized this limitation and constructed a novel 3D frame buffer with flexible scan-out circuitry for the specific purpose of studying display latency [RMRK99]. Although successful at reducing latency, their system lacked vertical parallax (i.e. it limited the viewing positions to points along a specific line), and it limited the image plane to the face of the display device. Moreover, it required off-line rendering to pre-compute the contents of the 3D frame buffer. The display stage of the PixelView architecture extends and generalizes Regan et al. into a 4D frame buffer. Its flexible scan-out stage supports both horizontal and vertical parallax. Furthermore, the PixelView architecture contains geometry and rasterization stages that allow primitives to be rendered into the 4D frame buffer.

Another active area of research has been the decoupling of slow rendering processes from the interactive demands of viewing and animation. Others have proposed special-purpose hardware [RP94] and software [MMB97; BWG03; Gla88] rendering systems to address this problem. Many of these systems also incorporate 3D [RMRK99] and 4D [War98; Bal99] frame buffers or ray caches, which are sampled and interpolated to produce view-specific renderings. Most of these systems operate as lazily evaluated caches, meaning that samples from previously rendered views are combined with samples from the current viewpoint. This approach generally requires no, or very little, variation in each point's reflectance as a function of viewpoint, with the notable exception of Bala [BWG03] who maintained a metric describing the range of views over which each radiance sample was valid.

Our approach renders the contribution of each primitive into all possible views. This affords a heretofore-unexploited type of coherency that is currently unavailable to traditional view-dependent rendering architectures, at the price of potentially rendering rays that might go unseen. There has even been some work on exploiting the coherence of rendering due to smooth variations in viewing position [Hal98]. This system effectively transformed and harnessed the power of 3D rendering to allow space-time or EPI rendering. We attempt to exploit the same sort of coherence in our shading approach. However, we do not focus on rendering epipolar planes one-at-a-time, but instead render the out-going radiance from each 3D point and use z-buffering to resolve visibility.

Our system relies on substantial preprocessing of display primitives much like Reyes [CCC87] and Talisman [TK96]. Specifically, we are able to render directly only those primitives that can be appropriately subdivided in world space. Moreover, as the average size of a rendering primitive shrinks, alternative primitives have been suggested. Examples of these include point-based models [RL00; PZvBG00; ZPKG02; PKKG03] and image-based models [SGHS98]. PixelView is capable of directly rendering and displaying point-based representations with view-dependent effects, as well as light fields [LH96] and lumigraphs [GGSC96]. Each display type can be easily combined with any of the others. Furthermore, the sampling and reconstruction strategies used in PixelView draw heavily on those developed for point-based primitives and light field rendering.

4. The PixelView Architecture

This section describes the various stages in the PixelView architecture. Note that this section is intended to provide a general, abstract description of the architecture, in contrast to the specific, concrete implementation presented in Section 5. Referring to Figure 2, we begin with the "lower half" of the pipeline (i.e. rasterization and display).

4.1. A 4D Frame Buffer

In PixelView, the standard 2D frame buffer is replaced with a 4D ray buffer. Frame buffers are commonly described as an array of pixels, but in the context of 3D rendering, they are more accurately characterized as an array of rays seen from a single viewpoint. This "frame buffer as ray buffer" concept is appropriate for both ray-casting and OpenGL-style renderers.

View independence is achieved by generalizing the 2D "ray buffer" into 4D. The resulting structure is, in essence, a light field/lumigraph, with rays specified by their intersection with two parallel planes [LH96; GGSC96]. Following the notation of Gortler et al. [GGSC96], we call these two planes the s-t plane and the u-v plane. Our frame buffer is thus a 4D collection of radiance values, a finite sampling of light rays parameterized by (s,t,u,v). Once the 4D frame buffer has been "populated", novel views can be generated without the need to re-render scene geometry.

4.2. Display/Scan-Out

New images can be created during scan-out by taking a 2D slice of the 4D frame buffer. This involves mapping scan-out pixel coordinates (i,j) into ray coordinates (s,t,u,v). Conceptually, the pixel coordinates are specified by rays, and the intersection of these rays with the s-t plane and u-v plane defines an (s,t,u,v) quadruple, as shown in Figure 3. The resulting mapping is given by the following four linear rational equations (a derivation is given in Appendix A).

$$s(i,j) = \frac{A_s i + B_s j + C_s}{A_w i + B_w j + C_w} \quad (1) \qquad t(i,j) = \frac{A_t i + B_t j + C_t}{A_w i + B_w j + C_w} \quad (2)$$

$$u(i,j) = \frac{A_u i + B_u j + C_u}{A_w i + B_w j + C_w} \quad (3) \qquad v(i,j) = \frac{A_v i + B_v j + C_v}{A_w i + B_w j + C_w} \quad (4)$$

The A, *B*, and *C* coefficients are defined with respect to the current position and orientation of the virtual camera (i.e. the current view). Each equation has a numerator of the same form, though the coefficients are different. The denominator is identical for all four equations.

These four equations represent the 2D planar slice of the 4D frame buffer, which maps an (i,j) pixel coordinate to an (s,t,u,v) ray index. As the scan-out iterates through i and j, these equations generate addresses into the 4D data structure. Note that the denominator varies with i and j, requiring per-pixel division. If, however, the image plane of the virtual camera is restricted such that it is always parallel to the s-t and u-v parameterization, the equations simplify to the following linear expressions.

$$s(i,j) = A'_s i + B'_s j + C'_s \quad (5) \qquad t(i,j) = A'_t i + B'_t j + C'_t \quad (6)$$

$$u(i,j) = A'_u i + B'_u j + C'_u \quad (7) \qquad v(i,j) = A'_v i + B'_v j + C'_v \quad (8)$$

Thus, we now have simple linear expressions in terms of pixel coordinates (i,j). Such expressions map well to hardware. The sacrifice for this elegance is that the orientation of our camera's image plane is now restricted. However, for applications such as autostereoscopic and CAVE-style virtual reality displays, it is practical to define the fixed viewing surface to be parallel to the s-t and u-v planes. That is, for these applications, this viewing configuration is inherent.

Our scan-out equations are similar to those used by Regan et al. in [RMRK99], but they are slightly more general. By limiting the viewing positions to points along a specific line, and by limiting the image plane to the face of the display device, Equations 5, 7, and 8 can be further simplified to those used in [RMRK99] (Equation 6 becomes unnecessary).

4.3. Point Casting and Scatter

Given a 4D frame buffer and the equations for scan-out, we must next tackle the issue of how to fill the frame buffer. The defining characteristic that separates the PixelView architecture from being "just a light-field viewer" is its ability to render geometric primitives into its 4D frame buffer. The geometry processing stage (i.e. subdivision and shader/texture) produces world-space points, each with an associated radiance map. These maps represent a sampled version of the outgoing radiance for each point. This radiance needs to be added to the 4D frame buffer for each s-t sample location. This process is dubbed "point casting" to indicate that a single point broadcasts its radiance out to a set of s-t sample points, instead of the more typical mapping to just a single camera's center of projection.

As shown is Figure 4, the process is performed by first iterating over the set of all s-t sample locations and finding the ray connecting the current sample location with the point primitive. This ray is then intersected with the u-v plane, and the 2D coordinate of that intersection determines the u-v sample location. This represents the (s,t,u,v) location in the 4D frame buffer where the radiance will be stored if it passes the 4D z-buffer test at that location. The

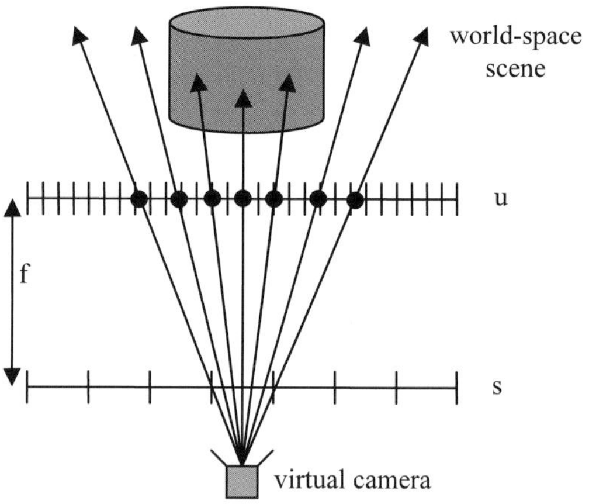

Figure 3: *2D depiction of rays displayed in scan-out. Each frame, PixelView evaluates a linear expression based on the camera matrix to determine which (s,t,u,v) rays are seen.*

intersection is given by the following two equations (a derivation is given in Appendix B).

$$u(s) = A''_u s + B''_u \quad (9) \qquad v(t) = A''_v t + B''_v \quad (10)$$

Thus, we must once again evaluate a linear expression, similar to the calculations required during scan-out. In this case, the coefficients are related to the (x,y,z) location of the scene point, the location of the u-v plane (f), and the terms in the matrices of Equation 16 (see Appendix B).

Note that, in general, the u-v intersection will not be an integer value. That is, the ray will usually not fall exactly in an (s,t,u,v) bin. Thus, the radiance contribution should be "scattered" into nearby bins.

4.4. Subdivision

Point primitives are ideal for the PixelView architecture because they specify a distinct ray to each s-t and u-v sample point, simplifying the point-casting process. However, it is possible to convert other primitives into this point-based representation through world-space subdivision. Note that all subdivision occurs without considering occlusion, which is handled through z-buffering at the time of point casting.

Polygonal models are subdivided into points until a sufficient density is achieved so that holes are not created when the points are mapped into the 4D frame buffer. The first pass of subdivision converts each polygon into a set of triangles to simplify subdivision. Subsequent passes perform a standard midpoint subdivision algorithm. The process continues until the length of each side of the triangle is less than half the size of a sample in the 4D frame buffer. This stopping criterion is not applied to the region as a whole, but instead to each individual subdivided triangle to allow for more precise sampling on large polygons. The method is similar to the Reyes approach [CCC87], but without explicit advance knowledge of the camera location, requiring uniform world-space subdivision.

Higher-order surfaces use subdivision algorithms more specific to their natural parameterizations. Thus, each of the supported higher-order surfaces is implemented with its own world-space subdivision procedure. For instance, a sphere is subdivided by algorithmically distributing points over the sphere's surface. By avoiding a traditional triangulation stage, the creation of geometry artifacts is avoided. Finding improved subdivision methods for primitives that are more complex remains an area for future research.

4.5. Shading

For point geometry fed directly into the hardware, PixelView needs color-shading information for each surface point's (x,y,z) coordinate. For polygonal and higher-order objects computed by PixelView, illumination must be applied.

Current rendering architectures depend on high-speed parallelism to accomplish the shading operations required at each point. However, the speed of a shading model is not as important when scan-out is decoupled from rendering/shading, as it is in PixelView. In order to create the illusion of smooth camera motion, traditional hardware must re-render an entire scene into a double buffer and then swap it at the next vertical refresh. This fill rate into the double buffer determines the frame rate of the system. Because PixelView can create user camera motion without resorting to reshading the entire scene, its frame rate is determined by the speed of the scan-out hardware, which is a constant independent of scene complexity. While the user is moving the view within a scene, a new scene is being shaded and is swapped in when complete. This independence allows implementation of more complex local and global shading algorithms with less concern for completing a frame by an arbitrary deadline, while still providing guaranteed view-update latency.

Hardware shading might be implemented using points output from subdivision and the normals generated during that process. For global illumination, this would imply a retained-mode graphics interface for primitive specification, or an immediate-mode interface for local illumination.

PixelView has the ability to interface with software renderers and offline storage of 4D frame buffers (e.g. 3D movies). This data could be loaded directly into the 4D frame buffer, bypassing PixelView's shading stage in a manner similar to how image-based rendering models are loaded. Another option is to transmit raw points and precomputed view-dependent global illumination, and then lets PixelView handle the placement of those points into the 4D frame buffer.

4.6. View-Dependency of Pixels

The ability to generate a 4D frame buffer without knowing the location of the user's camera in advance introduces the problem of accounting for scene elements with view-dependent specular reflections. Obviously, if all objects in the scene are diffuse, this is not an issue. If there is specular data, a method for shading, formatting, and point casting this extra information is needed.

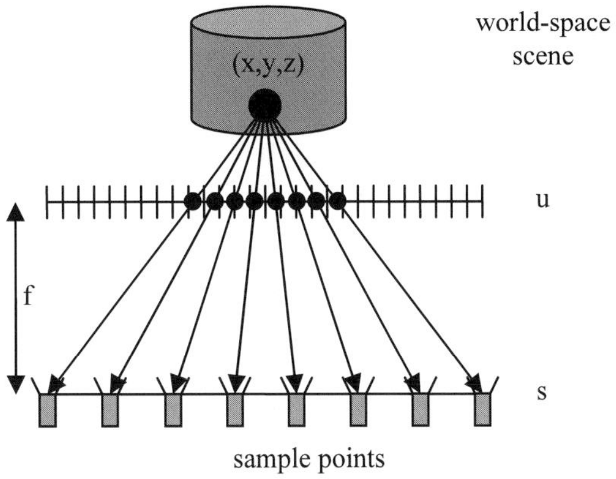

Figure 4: *2D depiction of a world-space point being rasterized into the 4D frame buffer via "point casting". The world-space point is tested against each s-t sample point to determine where u-v intersections occur. The point's radiance is then applied to those (s,t,u,v) samples.*

The mapping of each outgoing view direction to the s-t sample points will be referred to as the *radiance map* of a point. For example, Phong radiance maps tend to be simpler than ray-traced reflective maps, which can resemble point-specific environment maps.

Dealing with point-specific radiance maps raises the issue of computing large numbers of samples for each point and transmitting them to the hardware. Luckily, a great deal of exploitable coherence exists in the raw maps [LH96]. There is much more radiance-map coherence for true object points than for arbitrary points in space. This coherence allows radiance maps to be transferred to and within the PixelView hardware as compressed data. For reflection and refraction, the data is not as structured, but in cases with a limited viewable range, the variation in illumination is far smaller than that found, for example, in a spherical environment map. This data could still be compressed using a more typical compression algorithm, because it exhibits the same type of spatial coherence seen when viewing the world through a normal camera.

The general structure of point-specific radiance-map coherence implies that wavelet-based compression would be a natural fit. However, there exists the opportunity to create compression algorithms and basis functions that are better suited to a specific shading model, such as for specular highlights.

The point-casting process is similar for both diffuse and specular points. PixelView acknowledges the difference between diffuse and specular points by requiring only a single color for a diffuse point and a full radiance map for all s-t samples for a specular point.

5. A PixelView Implementation

In order to demonstrate the feasibility and advantages of the PixelView architecture, we have chosen to implement a proof-of-concept prototype (see Figure 1), with the following objectives:

- To examine the feasibility of dynamic view selection at the time of pixel scan-out
- To measure the memory bandwidth utilization of dynamic view selection and investigate the impact of different memory organizations on the required bandwidth
- To investigate the addressing, coherency, and fill-rate implications of illuminating and shading each primitive from multiple viewing directions
- To explore the algorithm simplifications and tradeoffs implied by a practical hardware implementation

Some of these objectives could have been addressed entirely in simulation. However, modern system design tools make it quite easy to move rapidly from a functional simulation model to a FPGA prototype. Moreover, the advantages of physical prototypes are that they do not wallpaper over many of the engineering issues that can be easily overlooked in simulation, such as accurate memory models, signal distribution, routing delays, floor planning, and datapath complexity.

Another possible implementation avenue for the PixelView architecture would be to map it onto one of today's Graphics Processing Units (GPUs). It is clear that the trend in graphics hardware is towards increasing flexibility and generality. In effect, GPUs can be viewed as programmable parallel-processing units. We expect this to become increasingly true in the future. However, there are many aspects of GPUs that, at present, are rendering architecture specific or unexposed. Examples of this include frame buffer scan-out logic and memory organization, both of which are critical to demonstrating the feasibility of PixelView.

We have chosen to prototype only a limited subset of the PixelView architecture. Specifically, our system implements only one "light slab" [LH96] of an all-views architecture. A complete all-views implementation would include at least 4 (and more likely 6) slabs to enclose a region of empty space. More slabs would be necessary in scenes with intervening occluders. Of course, there are tradeoffs between the range of actual view-independence and the frequency of re-rendering, and we plan to investigate those tradeoffs more in the future. Even so, a single light-slab implementation is still technically interesting, because it maps directly to a through-the-window viewing paradigm. Thus, it would be able to support interactions similar to those of a single-wall CAVE VR architecture, but with support for correct stereo viewing for multiple participants.

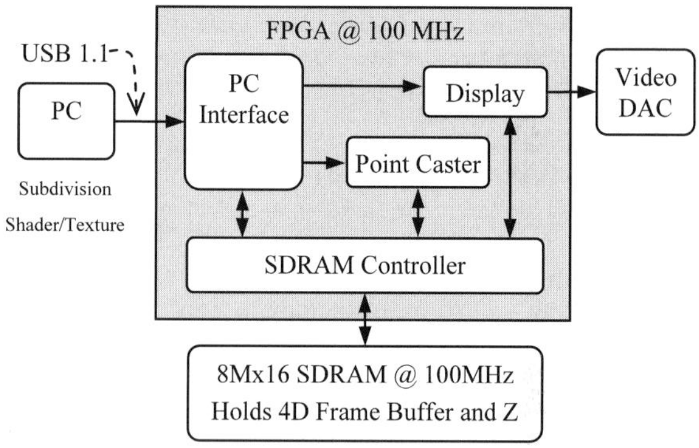

Figure 5: *Block diagram of the PixelView prototype. Point casting and display occur independently, allowing scan-out rate to be decoupled from rendering rate. A host PC implements the subdivision and shader/texture stages.*

Lastly, we have also chosen to use a general-purpose host processor to emulate the geometry processing stages of the PixelView pipeline. These stages generally require floating point computation and involve more decision-making and variations in control flow than the later stages of the PixelView pipeline. We intend to investigate a hardware version of this front-end geometry processing in the future.

5.1. Hardware Prototype

The hardware begins with the point-casting stage. Recall that the shader/texture stage outputs world-space (x,y,z) points along with their associated radiance maps that capture view-dependent reflectance. The hardware "rasterizes" these points into the 4D frame buffer. Visibility is determined via a corresponding 4D z-buffer. Scan-out is an independent function that uses view updates from the host PC to determine the latest view parameters at the start of each frame. This section gives details about how this hardware is implemented.

Figure 5 features a block diagram of the hardware system. The primary components are a single Xilinx XC2S300E Field Programmable Gate Array (FPGA), a single 8Mx16 SDRAM, a 10-bit video DAC, and a USB 1.1 port. The Xilinx and the SDRAM are both clocked at 100 MHz. The 4D frame buffer and 4D z-buffer reside in the SDRAM, each occupying 4Mx16. The frame buffer contains 16-bit RGB565 color values, and the z-buffer contains 16-bit two's complement depth values. Since we are limited to 4Mx16, the 4D buffers are organized as 8x8 s-t sample points, each with a corresponding set of 256x256 u-v samples.

We will first focus on point casting. The point caster receives (x,y,z,color) information from the PC. It uses this information to iterate over the s-t sample points and calculate the corresponding u-v bins. Each s-t sample point generates a conditional write to update the 4D frame buffer at the closest u-v sample. However, the write occurs only if the z value of the current point is "nearer" than the value currently at the corresponding location in the 4D z-buffer. Given enough (x,y,z) points, the 4D frame buffer will be "fully populated" and appear as a solid image.

Recall from Section 4.3 that determining u and v is a relatively straightforward evaluation of a linear expression. Specifically, it is an incremental calculation that utilizes accumulators. No multiplication or division is necessary.

Turning now to the scan-out/display hardware, it runs at a constant 640x480 at 60Hz. After vertical sync (i.e. at the start of each frame), the hardware grabs the latest view parameters from the PC. It then begins to iterate through (i,j) pixel coordinates just like a normal 2D scan-out. The difference is that the pixel coordinates for conventional scan-out map directly to frame buffer addresses, whereas our scan-out must calculate the equivalent (s,t,u,v) quadruple. The mechanisms for doing these calculations are similar to those for point casting, as they both involve the evaluation of a linear expression.

5.2. Scan-Out Memory Bandwidth

Bandwidth is always a primary concern in any graphics hardware system. It is reasonable to wonder how our 4D memory organization affects DRAM performance. Our hardware system has two main users of memory bandwidth: the point-casting stage and the display stage. The display stage generates sequential (i,j) pixel values as in normal scan-out, but these values are then mapped into 4D. Referring back to Figure 3, note that the s-t plane is sparsely sampled, whereas the u-v plane is finely sampled. Thus, scan-out tends to stay in a single s-t bin for several pixels. We can use this to our advantage by "swizzling" (s,t,u,v) such that the bank/row address of the DRAM stays in the same page for several iterations (Figure 6). Using this scheme, we achieve page-hit rates well over 90% during scan-out, and scan-out consumes only 4 percent of available bandwidth.

5.3. Point-Casting Performance

The prototype system is able to transmit and rasterize over 80,000 points per second. This equates to over five million rays per second written into the 4D frame buffer. Note that this rate is limited by the USB 1.1 interface between the host PC and the prototype, not by the point-casting hardware. At this rate, point casting consumes about 17 percent of available bandwidth. Factoring in scan-out and SDRAM auto-refresh, we have consumed less than 25 percent of available bandwidth. Thus, we have sufficient bandwidth to add additional point-casting units to process incoming points in parallel and increase point-casting performance. However, we would still ultimately be limited by the speed bottleneck of the USB interface, and so we chose not to do this.

6. Results

An important result of our research is the completion of the working prototype that demonstrates the benefits, capabilities, and plausibility of the PixelView architecture. This section provides examples of all supported primitive types captured from the PixelView prototype.

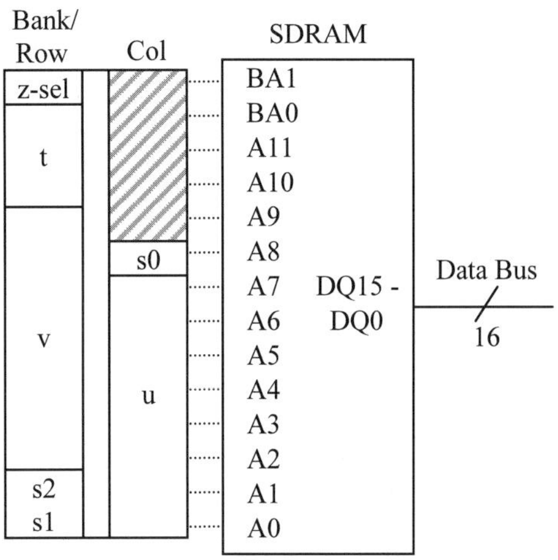

Figure 6: *Swizzling of (s,t,u,v) samples into SDRAM addresses to achieve increased scan-out coherence. Half the banks are used for z-buffer data and half for color data, determined by the high order bank pin.*

The example images are captured from a VGA scan converter connected to the VGA output of the PixelView prototype. All examples run at a constant frame rate of 60 frames per second within a 256x256 window inlaid in a 640x480 display. The images shown here have been enlarged to show detail. A better idea of the final output can be gained from watching the supporting video. The subdivision for these models has been computed on the host PC and transmitted to the board in a point stream format consistent with what a hardware-based subdivider would output.

The quality of reconstruction in these results is largely a function of memory size. The SDRAM can store only 8x8 s-t samples in the 4D frame buffer at a resolution of 256x256 and still allow room for 16-bit z-buffering. In addition, the reconstruction method used here is nearest neighbor. Better reconstruction methods, such as quadrilinear interpolation, could be implemented at the cost of additional memory bandwidth.

Our polygonal models are subdivided by the host PC and transmitted to the prototype as specular point samples. The model in Figure 8 shows a 3263-faced polygonal model of a cow with vertex normals. This is subdivided into 600899 Phong-shaded point samples, which are sufficient to render the scene for any view. The excess points are removed by z-buffering and all occlusions are properly resolved.

The point based terrain map data shown in Figure 9 represents a dataset of 2048x2048 (x,y,z) texture mapped points forming a model of Puget Sound in Washington. This data is easy for PixelView to work with because it does not need to subdivide or shade the points before point casting.

Image-based rendering datasets can be loaded directly into the 4D frame buffer and then viewed from a virtual camera position. The light field in Figure 10 was taken with a real-world light-field capture rig and digitized. The data set is comprised of 8x8 camera images, each with a resolution of 256x256 pixels, which map directly into the 4D frame buffer.

The final example, Figure 11, depicts 14 reflective and refractive spheres against a texture-mapped background; all described as higher-order surfaces. Ray tracing was used to compute the view-dependant radiance maps of each point in this data set. Each point and its radiance map is point cast and then scanned out at 60 frames per second.

7. Discussion and Future Work

While our prototype system addresses many of the issues posed by the PixelView architecture it leaves many unaddressed. Two issues of particular interest are how the PixelView architecture scales to a significantly larger 4D frame buffer, and how its rendering performance scales as multiple simultaneous views are supported. How to animate objects is another area of interest.

7.1. Scalability

We envision that a minimal practical PixelView system would incorporate at least 4G ray samples. Addressing such a large memory block poses immediate issues. Consider the cost of clearing the buffer's z and color values alone. We believe that the solution to this problem is to distribute the 4D frame buffer over multiple functional units, each with a slice of the 4D frame buffer that shares small overlap regions with its adjacent neighbors (Figure 7). We envision that the granularity of these modules would match the prevailing memory density, providing one geometry processing and rasterization/display unit per memory chip in order to maximize the system bandwidth.

However, this need for high memory bandwidth is further aggravated by the need to support multiple simultaneous views. If a PixelView-like system were employed to drive an autostereoscopic or holographic display, then it is conceivable that every ray in the frame buffer would need to be read out at the update rate. While the update rate might be minimal for a static scene, we would still like to support animations and dynamic scenes at interactive rates. We believe the solution here is redundancy. Each ray within the 4D frame buffer might need to be replicated as many as 16 times to support in excess of 1024 views. This further increases the memory requirements by a factor of 16 to 64G rays. While this seems unimaginable now, in ten years time it might be both achievable and affordable.

7.2. Animation

In a 4D frame buffer, the notion of a partial screen clear seems especially helpful, particularly for supporting animation. Ideally, one would like to remove all contributions from a given point set in a past frame and re-render only those portions of the frame buffer that have changed. An old double-buffering trick used in traditional systems might prove to be particularly useful for the PixelView system. Specifically, one can render static elements into a background buffer and then reload them to initialize each rendered frame. This trick is seldom used because it fixes the viewpoint in favor of fast animation, but it is particularly inviting for a view-independent architecture, like PixelView.

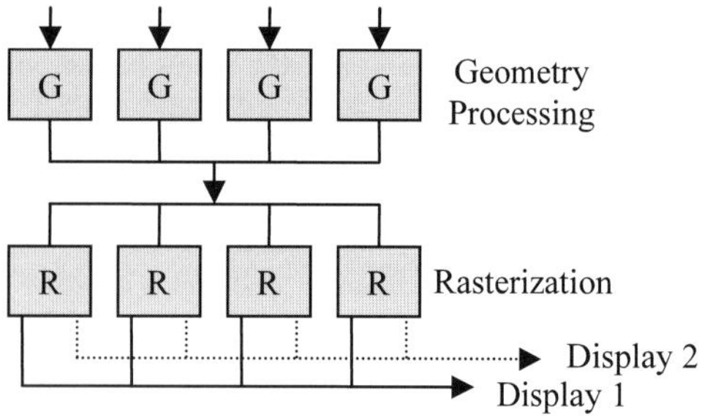

Figure 7: *Block diagram of scalability via multiple PixelView devices. Geometry processing and rasterization are performed in parallel, with each rasterizer controlling part of the light field, and subsequently part of each scan-out.*

7.3. Other Issues

A more general-purpose implementation of PixelView than our prototype would include the complete linear rational addressing implied by Equations 1 through 4. This seems achievable, because it is similar to the per-pixel divide required for perspective correct interpolation in today's hardware. We only require that similar circuitry be used in the frame buffer scan-out stage, where it could be heavily pipelined.

The image quality of PixelView could be improved with better reconstruction. A naïve implementation of quadrilinear interpolation causes an additional 16x increase in memory bandwidth. However, caching methods could be employed to reduce bandwidth requirements.

The point-casting bandwidth of PixelView could also be reduced considerably by including compression of the radiance maps. Off-the-shelf methods, such as wavelet compression, would probably do a good job, but we are particularly interested in the possibility of compression methods with basis functions optimized for radiance modeling. Any new compression bases would probably need to be used in combination with compression methods for general images, because the limiting case of a mirror reflector leads to a general image. Adding compression would increase computation in the primitive processing front end, but this increase is probably offset by the reduction in communication bandwidth.

8. Conclusions

We have presented PixelView, a new hardware rendering architecture that generates all possible views of a scene after a single traversal of the scene description. PixelView supports a wide range of rendering primitives, including polygonal meshes and higher-order surfaces. It also supports recently developed point and image-based rendering primitives. Moreover, we have developed a working hardware prototype of a 4D, z-buffered frame buffer that supports dynamic view selection at raster scan-out to demonstrate the viability of such a system. Graphics systems like PixelView are capable of supporting extremely low display update latencies. They will also enable efficient rendering of shared virtual environments supporting multiple simultaneous participants. In the future, view-independent graphics rendering architectures, like PixelView, will also be essential to provide the multitude of viewpoints required for real-time autostereoscopic and holographic display devices.

References

[Bal99] Bala, K., "Radiance interpolants for interactive scene editing and ray tracing" Doctorial Thesis, Dept. of Computer Science and Electrical Engineering, Massachusetts Institute of Technology (MIT), 1999.

[BWG03] Bala, K., Walter, B., and Greenberg, D., "Combining edges and points for interactive high-quality rendering", Proc. SIGGRAPH 2003.

[BFMZ94] Bishop, G., Fuchs, H., McMillan, L., and Zagier, E.S., "Frameless rendering: Double buffering considered harmful", Proc. SIGGRAPH 1994, pp. 175-176.

[CCC87] Cook, R. Carpenter, L, and Catmull, E., "The Reyes image rendering architecture," Proc. SIGGRAPH 1987, pp. 95 102.

[Gla88] Glassner, A.S., "Spacetime ray tracing for animation", IEEE Computer Graphics and Applications, 60-70, 1988.

[GGSC96] Gortler, S.J., Grzeszczuk, R., Szeliski, R., and Cohen, M. F., "The lumigraph", Proc. SIGGRAPH 1996, pp. 43-54.

[Hal98] Halle, M., "Multiple viewpoint rendering", Proc. SIGGRAPH 1998, pp. 243-254.

[LH96] Levoy, M. and Hanrahan, P., "Light field rendering", Proc. SIGGRAPH 1996, pp. 31-42.

[MMB97] Mark, W. R., McMillan, L., and Bishop G., "Post-rendering 3d warping", Proc. 1997 Symposium on Interactive 3D Graphics, pp. 7-16, 1997.

[OCMB95] Olano, M., Cohen, J., Mine, M., and Bishop, G., "Combating rendering latency", 1995 Symposium on Interactive 3D Graphics, 1995.

[OKTD02] Owens, J.D., Khailany, B., Towles, B., and Dally, W.J., "Comparing Reyes and OpenGL on a stream architecture," Proc. SIGGRAPH/ EUROGRAPHICS conference on graphics hardware, pp. 47-56, 2002.

[PKKG03] Pauly, M., Keiser, R., Kobbelt, L.P., and Gross, M., "Shape Modeling with Point-Sampled Geometry", Proc. SIGGRAPH 2003.

[PZvBG00] Pfister, H., Zwicker, M., van Baar, J., and Gross, M., "Surfels: Surface elements as rendering primitives", Proc. SIGGRAPH 2000, pp. 335-342.

[PBMH02] Purcell, T.J., Buck, I., Mark, W.R., and Hanrahan, P., "Ray tracing on programmable graphics hardware", Proc. SIGGRAPH 2002, pp. 703-712.

[RMRK99] Regan M., Miller G., Rubin S., and Kogelnik C., "A Real Time Low-Latency Hardware Light-Field Renderer", Proc. SIGGRAPH 1999, pp. 287-290.

[RP94] Regan M. and Pose R., "Priority rendering with a virtual reality address recalculation pipeline", Proc. SIGGRAPH 1994, pp. 155-162.

[RL00] Rusinkiewicz, S. and Levoy, M., "QSplat: A multiresolution point rendering system for large meshes", Proc. SIGGRAPH 2000.

[SGHS98] Shade, J., Gortler, S., He, L., and Szeliski, R., "Layered depth images," Proc. SIGGRAPH 1998, pp.231-242, 1998.

[TK96] Torborg, J. and Kajiya, J.T., "Talisman: Commondity realtime 3D graphics for the PC", Proc. SIGGRAPH 1996, pp. 353-363.

[War98] Ward, G.J., "The Holodeck: a parallel ray-caching rendering system", 2nd Eurographics Workshop on Parallel Graphics and Visualization, 1998.

[ZPKG02] Zwicker, M., Pauly, M., Knoll, O., and Gross, M., "Pointshop 3D: an interactive system for point-based surface editing, Proc. SIGGRAPH 2002, pp. 322-329.

Appendix A. Derivation of Scan-Out Equations

Equation 11 below gives a model of the "image plane" of the virtual camera used to specify the current view. Specifically, given a point $\boldsymbol{E}$ (the eye point), a vector $\boldsymbol{P}$ that spans the image plane in the i-direction, a vector $\boldsymbol{Q}$ that spans the image plane in the j-direction, and a vector $\boldsymbol{O}$ from the eye point to the (0,0) pixel of the image plane, Equation 11 maps pixel coordinates (i,j) to (x,y,z) points in world space.

$$\begin{bmatrix} x \\ y \\ z \end{bmatrix} = \begin{bmatrix} E_x \\ E_y \\ E_z \end{bmatrix} + \lambda \cdot \begin{bmatrix} P_x & Q_x & O_x \\ P_y & Q_y & O_y \\ P_z & Q_z & O_z \end{bmatrix} \cdot \begin{bmatrix} i \\ j \\ 1 \end{bmatrix} \qquad (11)$$

The following two equations represent a similar mapping for the s-t and u-v planes. Specifically, they define the world-space points that fall on the planes. In Equation 12, vector $\boldsymbol{L}$ spans the s-t plane in the s-direction, vector $\boldsymbol{M}$ spans the s-t plane in the t-direction, and vector $\boldsymbol{S}$ goes from the world origin to (s,t) = (0,0). Similar definitions exist for Equation 13.

$$\begin{bmatrix} x \\ y \\ z \end{bmatrix} = \begin{bmatrix} L_x & M_x & S_x \\ L_y & M_y & S_y \\ L_z & M_z & S_z \end{bmatrix} \cdot \begin{bmatrix} s \\ t \\ 1 \end{bmatrix} \quad (12) \qquad \begin{bmatrix} x \\ y \\ z \end{bmatrix} = \begin{bmatrix} F_x & G_x & U_x \\ F_y & G_y & U_y \\ F_z & G_z & U_z \end{bmatrix} \cdot \begin{bmatrix} u \\ v \\ 1 \end{bmatrix} \quad (13)$$

If we place the s-t plane at $z = 0$, define ***L*** parallel to the x-axis, and define ***M*** parallel to the y-axis, Equation 12 simplifies to Equation 14. Similarly, if we place the u-v plane at $z = f$, define ***F*** parallel to the x-axis, and define ***G*** parallel to the y-axis, Equation 13 simplifies to Equation 15.

$$\begin{bmatrix} x \\ y \\ z \end{bmatrix} = \begin{bmatrix} L_x & 0 & S_x \\ 0 & M_y & S_y \\ 0 & 0 & 0 \end{bmatrix} \cdot \begin{bmatrix} s \\ t \\ 1 \end{bmatrix} \quad (14) \qquad \begin{bmatrix} x \\ y \\ z \end{bmatrix} = \begin{bmatrix} F_x & 0 & U_x \\ 0 & G_y & U_y \\ 0 & 0 & f \end{bmatrix} \cdot \begin{bmatrix} u \\ v \\ 1 \end{bmatrix} \quad (15)$$

Setting the right side of Equation 11 equal to the right side of Equation 14, we can solve for s and t in terms of i and j. Likewise, we can solve for u and v in terms of i and j using Equations 11 and 15. This yields Equations 1 through 4 in Section 4.2.

Moreover, if the image plane of the virtual camera is restricted such that it is always parallel to the s-t and u-v planes, the P_z and Q_z terms in Equation 11 become zero, and consequently, the denominator becomes a constant. This results in Equations 5 through 8 in Section 4.2.

Appendix B. Derivation of Point-Casting Equations

Assuming the simplified s-t and u-v configuration from Appendix A, the following equation represents the ray starting at a particular s-t sample point and passing through a particular point on the u-v plane:

$$\begin{bmatrix} x \\ y \\ z \end{bmatrix} = \begin{bmatrix} L_x & 0 & S_x \\ 0 & M_y & S_y \\ 0 & 0 & 0 \end{bmatrix} \cdot \begin{bmatrix} s \\ t \\ 1 \end{bmatrix} + \tau \left(\begin{bmatrix} F_x & 0 & U_x \\ 0 & G_y & U_y \\ 0 & 0 & f \end{bmatrix} \cdot \begin{bmatrix} u \\ v \\ 1 \end{bmatrix} - \begin{bmatrix} L_x & 0 & S_x \\ 0 & M_y & S_y \\ 0 & 0 & 0 \end{bmatrix} \cdot \begin{bmatrix} s \\ t \\ 1 \end{bmatrix} \right) \quad (16)$$

Viewing this expression as a system of three equations, note that the last equation is simply $z = \tau \cdot f$. Consequently, you can easily solve for τ and substitute back to solve for u and v in terms of s and t. This yields Equations 9 and 10 in Section 4.3.

Figure 8: *Screen capture of a subdivided polygonal model.*

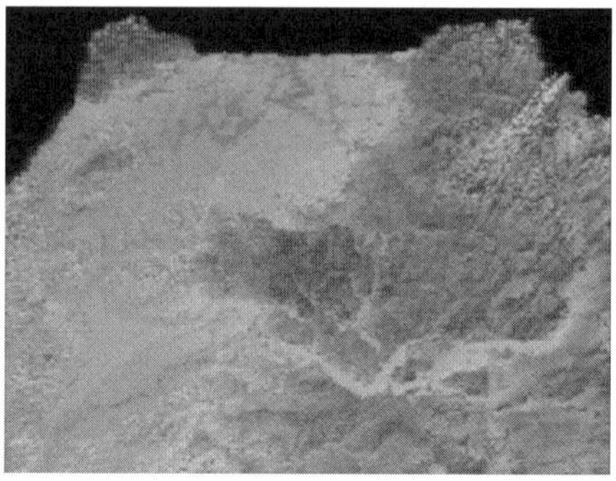

Figure 9: *4-million point-sample model of Puget Sound.*

Figure 10: *A "real-life" 8x8 by 256x256 light field.*

Figure 11: *Ray-traced scene with view-dependent points.*

Graphics Hardware (2004)
T. Akenine-Möller, M. McCool (Editors)

A Flexible Simulation Framework for Graphics Architectures

J. W. Sheaffer, D. Luebke, and K. Skadron

Department of Computer Science, The University of Virginia

Abstract

In this paper we describe a multipurpose tool for analysis of the performance characteristics of computer graphics hardware and software. We are developing Qsilver, *a highly configurable micro-architectural simulator of the GPU that uses the* Chromium *system's ability to intercept and redirect an OpenGL stream. The simulator produces an annotated trace of graphics commands using Chromium, then runs the trace through a cycle-timer model to evaluate time-dependent behaviors of the various functional units. We demonstrate the use of Qsilver on a simple hypothetical architecture to analyze performance bottlenecks, to explore new GPU microarchitectures, and to model power and leakage properties. One innovation we explore is the use of dynamic voltage scaling across multiple clock domains to achieve significant energy savings at almost negligible performance cost. Finally, we discuss how other architectural features and experiments might be incorporated into the Qsilver framework.*

Categories and Subject Descriptors (according to ACM CCS): I.3.1 [Computer Graphics]: Hardware Architecture

1. Introduction

Simulation has long been a vital tool for the study and design of computer architecture, but simulation of graphics architecture presents some unique challenges. Commodity graphics hardware is evolving at a tremendous rate, with each successive generation adding not only performance but fundamentally new functionality. In the time required to build a complex simulation infrastructure, the simulated architecture can easily become obsolete. Furthermore, the architecture of modern graphics processors or *GPUs* are largely secret; vendors in the highly competitive PC graphics arena are reluctant to release architectural details, such as texture cache design, even to their registered game developers. Most GPU simulators today can be broadly classified as cycle-accurate circuit-level simulations, which are extremely costly to run and available only to vendors because of the enormous manpower they require to produce, and functional emulators, which model the user-level interface (for example, a shading assembly language or API) without any underlying architectural structure. This situation, coupled with the sheer complexity and performance of modern GPUs, presents a significant challenge for researchers interested in exploring architectural innovations, modeling fine-grained effects such as intra-frame performance bottlenecks, or simulating power, leakage, and thermal properties of GPUs.

We present a simple, flexible simulation framework for graphics architectures that fills the gap between cycle-accurate VHDL/Verilog models and functional emulators. Our framework builds on the *Chromium* system [7], which intercepts and processes streams of OpenGL calls. We use Chromium to record a trace of graphics instructions during an application and then to instrument the rendering pipeline to accumulate statistics during playback of that trace. The instrumented trace is then run through the simulator itself, which is essentially a queue-driven, cycle-timer model of data and computation flow through the functional stages, decoupled by queues and caches, of the GPU. The resulting simulation framework is fine-grained enough to be useful for a variety of tasks – for example, we demonstrate power modeling of a hypothetical architecture – but efficient enough to analyze long (multi-second) runs of real-world applications. The framework is also flexible, allowing greater or lesser amounts of detail to be modeled and providing a natural structure for adding experimental pipeline stages or functional units via Chromium *stream processing units* or

SPUs. We call our fast Chromium-based simulation framework *Qsilver*.

To demonstrate the applicability of Qsilver, we present two simple studies. First, we analyze the performance of a simple hypothetical architecture running the recent Splash Damage game "Return to Castle Wolfenstein: Enemy Territory." We show that, as might be expected, different stages of the GPU pipeline act as the bottleneck during different frames and during different segments of a single frame. Next, we build a power model of the hypothetical architecture and use Qsilver to explore power-related architectural optimization.

Note that these studies serve as illustrative examples rather than serious architectural experiments. Our hypothetical GPU pipeline is a prototype, modeled only coarsely, and our power model represents a best-guess estimate of a current graphics processor. We emphasize that our goal here is not to perform conclusive architectural research but to describe a useful tool and demonstrate the sort of interesting studies that vendors and researchers, armed with the architectural details that our prototype model lacks, could easily perform with the Qsilver approach.

2. Related work and motivation

The advent of detailed but flexible, configurable, cycle-accurate CPU simulators in the 1990s for complex, superscalar architectures served as the catalyst for an explosion of quantitative research in the computer architecture community. The most prevalent simulator in academic architectural research is SimpleScalar [3]; other simulators used in specific circumstances include Rsim [6] for multiprocessors, as well as Simics [10] and SimOS [13] for capturing operating-system and multi-programmed behavior.

By describing instruction flow at the granularity of individual steps through the CPU pipeline, these simulators allowed research and design to move beyond simplified, imprecise analytical models or cumbersome, logic-level models. Instead, architects could analyze detailed tradeoffs under realistic workloads and estimate how various microarchitectural choices affected instruction throughput. Examples include the impact of different cache and branch predictor algorithms, the impact of different superscalar out-of-order instruction-issue designs, and the ability to evaluate a host of novel CPU organizations such as hyper-threading.

Subsequent power-modeling capability launched another round of innovation by allowing architects to estimate the energy efficiency of different processor organizations, verify that new microarchitecture innovations are justifiable from an energy-efficiency standpoint, and explore microarchitectural techniques for managing energy efficiency. The dominant power model today is Wattch [4], which uses calibrated analytical models to allow flexible and configurable estimation of power for a variety of structures, structure sizes, organizations, and semiconductor technology generations or "nodes". Other power models that use circuit-extracted data have been described, but they are based on a specific implementation and tend to be inflexible, especially in terms of scaling to future semiconductor technology nodes. These two approaches can be combined, using the circuit-extracted model as calibration for Wattch's analytical models; see for example a recent study of hyper-threading using IBM's circuit-extracted PowerTimer [9].

Our goals with Qsilver are to stimulate the same kind of innovation in the GPU community, to stimulate greater cross-fertilization with the general-purpose CPU architecture community, and to enable new studies in power-aware and eventually thermal-aware design. The purpose of this paper is to report results with a prototype to show the value of this kind of simulation approach.

3. A framework for simulation of graphics architecture

Driving our simulator is a trace of graphics commands, instrumented with additional statistics that describe the behavior of those commands. For example, in our system the graphics command that produces a triangle to be rasterized would be annotated with the total number of fragments generated, number of fragments written to the depth buffer, average number of texels accessed per fragment, and so on. Exactly which statistics must be gathered depends on the level of architectural detail we wish to model. We then feed this instrumented trace into a "cycle-timer" model that simulates the flow of data and computation through each stage in the decoupled microarchitecture of the GPU. From the cycle-timer model we can count the number of operations and estimate the computational load in each unit that we model, which in turn provides a basis for modeling power dissipation. We can also examine high-level behavior such as the migration of performance bottlenecks (vertex processing, fragment processing, memory bandwidth, etc.) between or within frames.

3.1. Capturing the trace

We use Chromium [7] both to capture the original trace and to gather the statistics used to instrument it. Chromium is a system for manipulating streams of OpenGL commands; common uses include splitting rendering across multiple machines for tiled display or implementing sort-first and sort-last parallel graphics architectures on clusters of PCs. The use of Chromium simplifies the task of capturing the behavior of real-world applications, since Chromium can be applied non-invasively to applications for which source code is not available and can store an application's OpenGL stream to disk for playback and reproducible analysis later. Chromium also lets us avoid the development and running-time overhead of implementing simulations of low-level structures, such as rasterization or texture filtering hardware.

Instead we use OpenGL itself as a rasterization engine, and instrument the OpenGL stream using mechanisms such as occlusion queries (see http://oss.sgi.com/projects/ogl-sample/registry/ARB/occlusion_query.txt) and programmable shaders to gather the essential statistics. For example, we employ a Chromium SPU that renders each triangle twice, once into the destination image buffer and once into a scratch buffer in which the depth test is disabled. Occlusion queries wrapped around both triangles provide a count of fragments drawn and fragments created, which we later use for estimating computational load and memory bandwidth used by the fragment processor and the depth buffer.

Specifically, our prototype simulator implements the following SPUs:

Expand vertex arrays. For efficient rendering, OpenGL applications generally collect object geometry into vertex arrays. To simplify the task of associating fragments with the primitives (vertices and triangles) that produce them, we dereference vertex arrays into immediate-mode `glBegin`, `glVertex`, `glEnd` calls. Note that we can easily retain the vertex array organization in our annotations if, for example, we wish to accurately simulate the post-transform vertex cache (but our current prototype simply uses a statistical model of cache hits).

Triangulate complex geometries. We wish to count total versus occluded fragments generated by each primitive, but complex primitives such as triangle strips can self-occlude. In this SPU we split triangle strips, triangle fans, and complex polygons into triangles so that our occlusion queries are guaranteed to measure only planar polygons. Though our prototype implementation does not do so, this SPU could also generate triangles for other primitives such as thick lines and point sprites.

Unfold display lists. We also store display lists and play them back when referenced in the OpenGL stream. In addition to potentially containing self-occluding geometry, display lists may encapsulate changes to rendering state. These must be exposed to the simulator since we wish to track information such as texture accesses, which are affected by the number of textures bound, MIP-map filtering mode, etc.

Count visible fragments rasterized. The aforementioned SPUs condition the OpenGL stream for annotation; this SPU begins the actual data collection. As each triangle is rendered, we enclose it in an occlusion query, which returns the number of fragments that pass the depth test. We store this information in the annotated trace that will later be used to drive the cycle-timer model.

Count total fragments rasterized. To get an accurate estimate of depth-buffer bandwidth, we must also count total fragments rasterized. We do this by rendering each polygon again, this time into a separate buffer with the depth test disabled. Again, an occlusion query returns the total number of fragments rasterized.

Count average texture accesses per fragment. The number of texture accesses depends on several things: how many textures are bound, what form of texture filtering is enabled, and the per-pixel texture LOD of the fragments. Full trilinear MIP-mapping presents a challenge because the number of texels read for magnification (4) is different from the number read for minification (8). We could use a software rasterizer that directly evaluates per-fragment texture LOD, but this sacrifices much of the simplicity and speed that we gain by using Chromium and OpenGL to perform rasterization and filtering. Instead, we obtain an average texture access count by binding a texture that encodes the MIP level directly, then binding a fragment program that kills fragments based on the hardware-filtered texture value. Another occlusion query counts surviving fragments and uses the result to calculate average texture accesses per fragment.

In practice, the last three SPUs can be combined to increase rendering efficiency. In our experiments with the *Enemy Territory* game, the entire process of annotating a prerecorded OpenGL trace, preparing input for the simulator core, runs at near-interactive rates, roughly 3-15 frames per

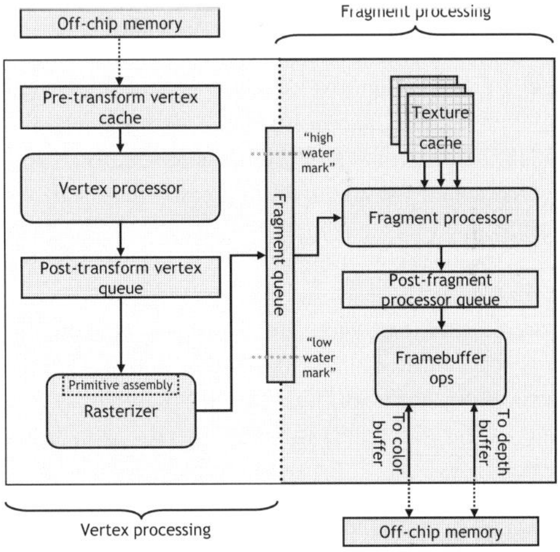

Figure 1: *The functional units and data flow of our hypothetical graphics architecture. The units can be classified as queues, caches, or data processing components. The fragment queue separates the vertex- and pixel-processing components of the GPU; we exploit this decoupling in the multiple-clock-domain experiment of Section 5.2.2.*

3.2. Applying the cycle-timer model

The instrumented trace consists of a stream of geometric primitives annotated with statistics denoting fragments produced and occluded, texture accesses, and potentially other information (such as vertex indices for post-transform cache analysis), as well as the relevant graphics state (such as number of textures, number/kind of lights, and texture filtering state) for each batch of geometry. Once the trace is acquired we next run a cycle-timer simulation that pulls the fragments and vertices, cycle by cycle, through each stage of the GPU pipeline. The pipeline is modeled as a series of functional units connected by queues and caches; Figure 1 shows our prototype model of a hypothetical fixed-function GPU.

In the cycle-timer model, Qsilver repeatedly advances a global time counter by one cycle and advances the simulation stage-by-stage backwards through the pipeline. For example, each cycle a maximum number of fragments may be written to the framebuffer (possibly limited further by memory bandwidth to the color and/or depth buffers). This increments operation counters in the "Framebuffer ops" unit (representing arithmetic operations performed for blending, Z-buffer comparison, etc) and drains the queue between that unit and the fragment processor. If the fragment processor is ready to produce a fragment (actually the fragment processor operates on tiles, which are 2×2 fragments in our model), the processed fragments are added to the queue. The rate of the fragment processor depends on the computation demanded by the rendering state (e.g., how many textures to interpolate), on the annotated trace statistics for that set of fragments (e.g., how many texture value interpolations are necessary due to magnification/minification), on the intrinsic architectural parameters of the fragment processor (how many pipelines, how deeply pipelined), and on the bandwidth to texture memory (accounting for texture cache behavior, which we currently model only statistically). Note that in a given cycle, the fragment processor may stall for several reasons: the subsequent queue may be full (this generally implies the system is framebuffer bound), because the texture fetches cannot be performed (system may be texture bound, or poor texture cache locality), or because the incoming queue is empty (vertex or rasterization bound). As the fragment processor produces fragments we increment operation counters that reflect the arithmetic operations incurred per fragment.

Qsilver follows a similar process through the various stages of the cycle-timer model. The fragment processor drains the "fragment queue", a queue of fragment interpolants (again, organized into tiles) filled by the rasterizer. This queue effectively partitions the GPU into vertex- and pixel-processing components; since a full or empty fragment queue implies that the vertex or fragment processor is stalled, queue occupancy provides a metric for whether the system is currently transform- or fill-rate limited. We therefore use occupancy to drive the multiple-clock domain experiment described in Section 5.2.2.

The rasterizer, which in our prototype incorporates primitive assembly, fills the fragment queue and drains the post-transform cache. Computational operations executed by the rasterizer are derived from the number of fragments generated per primitive (available from our annotated trace) and the number of active interpolants (dictated by current rendering state). The post-transform cache is modeled as a simple FIFO, which accurately reflects current architectures, but we do not currently model cache hits for indexed vertex arrays. Instead, the post-transform cache is treated as a queue, filled by the vertex processor.

The vertex processor (modeled as two separate pipelines in our hypothetical architecture) performs significantly more arithmetic per operation than other units. Again, the details depend on rendering state (e.g., number and kind of lights). To estimate the computational load of transforming and lighting vertices, we use a collection of compiled assembly code for implementing a fixed-function pipeline on programmable GPUs, obtained from a Cg model provided by Nvidia online (http://developer.nvidia.com/object/cg_fixed_function.html). This code includes the various permutations of point, directional, and spot lights, local- versus infinite-viewer lighting, and so on; we simply scale the appropriate operations by the number of lights and count the total operations. The vertex processor drains a pre-transform cache of vertices, which we approximate as a perfect (no-miss) cache in our prototype model.

To summarize, Qsilver runs a cycle-timer model that steps cycle-by-cycle; during each cycle every functional unit either advances in its computation, possibly producing an output for the next stage and incurring the cost of that computation's operations, or stalls as it waits on a queue or cache. By analyzing the activity of the various stages of the cycle-timer model over time, we can study the performance and bottlenecks of the system; by augmenting this information with a power model we can examine the power characteristics of the system. The simulation advances relatively quickly by architectural simulation standards; a simulated frame in our Enemy Territories study requires approximately 10-15 seconds to process on a 1.7GHz AMD Athlon with 512 MB RAM. This enables rapid exploration of design parameters such as the studies presented in Section 5.

4. Modeling power

To model power in Qsilver, we estimate the power for primitive operations like cache accesses, queue accesses, floating-point operations, register accesses, etc. Then, by identifying the primitive operations comprising a functional block in our pipeline, we can construct a power estimate for it.

To estimate the power for each primitive operation, we extract the cost of similar operations from an industrial, architecture-level power model to which we have access. This model is based on circuit-extracted data for a 180 nm high-performance superscalar microprocessor. We then scaled these estimates to match the structure sizes and bit-widths used in Qsilver, and further scaled them to the appropriate technology node according to CV^2f (power is proportional to CV^2f, where C is capacitance, V is voltage, and f is frequency. All of these scale with each successive generation of semiconductor technology). In our case, the current Qsilver prototype resembles an Nvidia GeForce 4 (but tuned to drive low framebuffer resolutions–see below), so we have chosen to model a 150 nm implementation running at 1.8 V and 300 MHz. Although there will certainly be differences in circuit-design style between a high-performance CPU and our GPU, the relative power cost among different operations in the high-performance superscalar CPU is likely to be a reasonable indicator of relative cost in the GPU. We also cross-checked the relative power dissipated in each primitive against both Wattch and a second industrial model; of course, these too are models for high-performance CPUs.

Although our current power-modeling approach is admittedly crude, it enables us to construct a prototype that demonstrates the potential benefits of modeling power in the GPU. Even though the prototype power model is clearly imprecise and not suited for detailed modeling, it allows us to explore basic energy-efficiency tradeoffs like the benefits of a more aggressive vertex or fragment engine, the ability of runtime techniques to reduce power in response to time-varying application behavior, and the affect of design tradeoffs on leakage power as it grows in importance with future technology generations.

The structures in the GPU fall into four major categories: cache, queue, register, and arithmetic. For the cache, queue, and register structures, after scaling the power value extracted from the reference model to account for CV^2f, we further scale it to account for the width, height, and number of ports in the GPU. For purposes of our prototype, we scale linearly with each dimension. For example, our reference model uses a direct-mapped, 64 KB, single-ported cache, while our texture cache has four ports. After scaling for technology, we therefore account for the extra ports by multiplying by four.

The data-processing units—for example, the vertex transform-and-light unit—are less straightforward. We deal with these units by assuming they are driven either by microcode or a finite state machine and determine the primitive operations they execute, such as dot products, multiply-adds, etc—for this, we use the collection of compiled assembly code for the model of the fixed-function vertex processor. We then further break these down into basic floating-point multiplies and adds wherever possible (the main exception being transcendental functions). In the case of the vertex transform-and-light unit, we assume that because these operations are so complex, it is not pipelined, but rather it consists of a floating-point multiply-add array and that each microcode operation makes one pass through this hardware. (This assumption is also the basis for our model of how many cycles it takes to traverse the transform-and-light stage.) Our reference power model is also based on a floating-point multiply-add array (double-precision in this case), so after scaling to account for CV^2f, we can simply scale the resulting power value to account for the number of bits involved in each microcoded step. We currently scale linearly with the number of bits in each operation, although this oversimplifies the actual power scaling. In addition to the cost of the arithmetic operations, we also account for the power dissipated in the microcode memory and in reading and writing the register file.

Finally, we augmented the power estimate for each block to include power due to leakage currents. Leakage current arises from the fact that in deep-submicron technologies, transistors are never fully cut off. At 150 nm, leakage already accounts for approximately 10% of total power dissipation (this assumes an operating temperature of approximately 100°C, since leakage is exponentially dependent on temperature). Non-ideal device scaling means that leakage is growing proportionally worse with each successive technology generation; the International Technology Roadmap for Semiconductors (ITRS) [18] projects that it will grow to 50% or more within a few years as we reach the 70 nm node. This is important because it means that units which are clock gated still dissipate substantial power, and so trying to improve energy efficiency by reducing switching activity and/or improving clock gating will be less helpful in future generations, while optimizations that put blocks to sleep (like disconnecting them from their voltage supply using sleep transistors) or reduce the operating voltage (as in the multiple-clock-domain results presented in Section 5.2.2, become more helpful.

Following the approach used in Wattch, we model leakage as a simple fraction of the unconstrained switching power. When a unit is idle and clock gated, instead of dissipating its associated dynamic power that has been derived from our reference model, it dissipates a technology-dependent fraction thereof. The appropriate leakage ratio is taken from ITRS, e.g., 10% for 150nm and 50% for 70nm.

Again, this power model is imprecise and can only serve as a prototype. We have made many guesses or simplifying assumptions about the internal microarchitecture of the GPU, unit latencies and power dissipation, and so forth; and our power estimates are derived from data for a high-performance microprocessor that may use different circuit-design styles than would be employed for a GPU. Just as developing a more detailed and accurate performance model is an important area for future work, so is developing a detailed and accurate power model.

Validating the current prototype model is difficult, because we have no reference against which to compare. We can however perform some sanity checks. First, we verified that power dissipation in each block tracks its activity factor. Next, we looked at the power density in each block. The two blocks dissipating the most power are the vertex engine and texture cache, both dissipating about 12 W at peak. Based on the rough GeForce 4, 12mm $\times$ 12mm floorplan in Figure 2 [17], this corresponds to about 0.7 W/mm^2, which is reasonable. Though we might expect each block to have a similar peak power density, the fragment engine only dissipates about 4.5 W. This may seem surprising given that the fragment processor is typically considered a computational "hot spot", but we have tuned our theoretical pipeline for a relatively low-resolution 800$\times$600 display–more appropriate for a game console architecture than the PC graphics chip we are using as a guiding floorplan. The lower resolution leads to a correspondingly lower overall arithmetic intensity, thus the apparent anomaly in power density. Similarly, the rasterizer only dissipates about 2.5 W. This is because our model of rasterizer operations does not include some computations: edge setup, primitive assembly, or clipping. Having verified that the relative power dissipation in each block is reasonable given the actions we are modeling, we also verified that total power dissipation, including blocks and activities we do not yet model, will be within a range of about 50-75 W.

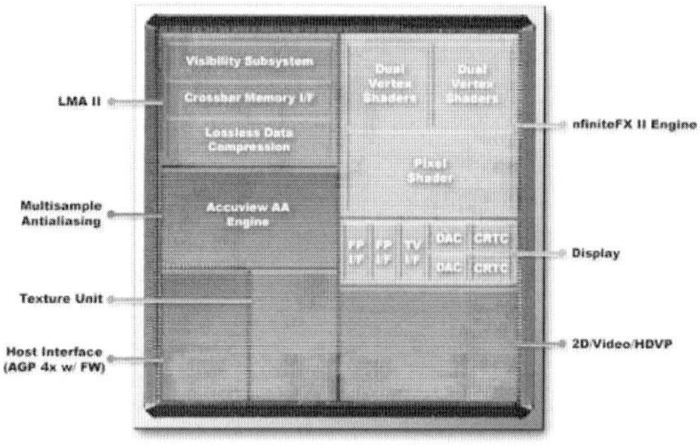

Figure 2: *GeForce 4 die photo from [17]. Note that the functional blocks in this marketing image are likely incorrect.*

5. Experiments and results

5.1. Performance analysis and experiment

One application of Qsilver is to analyze performance of a real application on a particular (perhaps hypothetical) architecture, with very fine granularity, enough to analyze intra- as well as inter-frame phenomena. For example, Figure 6 illustrates the evolution of a typical frame from *Enemy Territory*, along with timelines of some statistics from our model. Analysis of this timeline reveals interesting bottlenecks and behaviors. Images (a-d) are partially rendered frames at a certain cycle count; the ability to easily record such frames is a nice byproduct of using Chromium. In (a), the game is rendering a skybox using multiple passes, some textured and others untextured. As could be expected, the bottleneck is the fragment engine; vertex throughput is quite low (though notice that the textured pass uses additional polygons, perhaps so that fast-reject triangle culling will reduce rasterization load). In (b), the system has turned to rendering small details of trees and buildings; the vertex engine is now plateaued at maximum capacity and the game is transform limited. Shortly afterwards, (c) shows a sudden shift: the game is now rendering a road surface that contains few polygons, occupies much of the screen, and uses a MIP-mapped texture. Finally, in (d) the game is finishing the small head in the lower-left corner, which comprises many very small polygons and is thus heavily vertex-bound.

Of course, Qsilver can also be applied to the traditional task of microarchitectural simulators: performance studies. In the next section we describe a performance and power study based on varying the throughput of the fragment and vertex engines. One could also analyze performance while varying parameters such as the texture cache, queue lengths, memory architecture, and so on.

5.2. Experiment: power-aware graphics architecture

5.2.1. Energy-efficiency tradeoffs

To illustrate the value of Qsilver for power-aware design, we conduct two simple experiments in which we vary the throughput of the fragment and vertex engines to find the highest-performance and most energy-efficient design points (which are typically not the same points). Figure 3 plots the energy-delay-squared (ED^2) product [20], energy (E), and performance (frame rendering time, T) as a function of fragment processing rate. Figure 4 plots the same metrics as a function of vertex processing rate. All three metrics have been normalized to the results obtained with our default configuration: 5 cycles per tile for the fragment engine and 38 cycles per vertex for the vertex engine (no lighting is used in this experiment).

Clearly, the faster the engine, the better the throughput, but throughput reaches a point of diminishing returns at 4 cycles per tile in the fragment engine, at which point the power required to further improve performance is disproportionate to the miniscule marginal benefit. This makes 4 cycles/tile the energy-efficiency optimum, as shown by the ED^2 product. Note that this optimum differs slightly from the design point we initially chose, as might be expected in an early prototype model.

A similar trend is observed for vertex rate, where diminishing returns in terms of performance are not clearly present, but energy begins to rise disproportionately at the most aggressive throughput settings. The ED^2 optimum is therefore at 5-6 cycles/vertex.

These qualitative trends in these results are fairly insensitive to our power modeling assumptions, shifting the curves up or down on an absolute scale but leaving them essentially unchanged once normalized.

ED^2 has been established by the low-power design community as a sound figure of merit for trading off energy

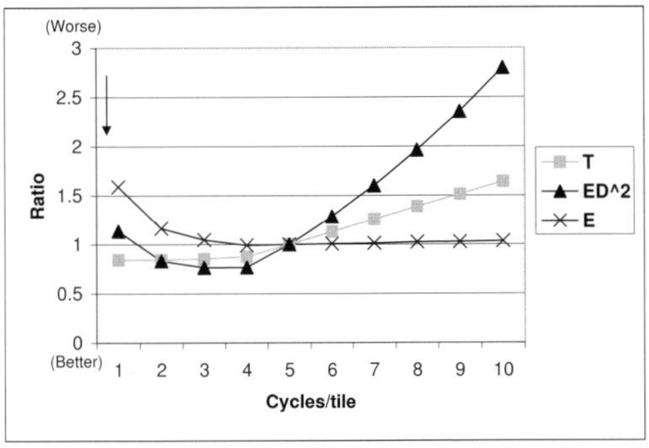

Figure 3: *Performance (T) and energy-efficiency data (ED^2 and E) for different fragment-processing rates. All results are normalized to our base case, which is 5 cycles/tile. Note also that in all cases, a smaller ratio is better.*

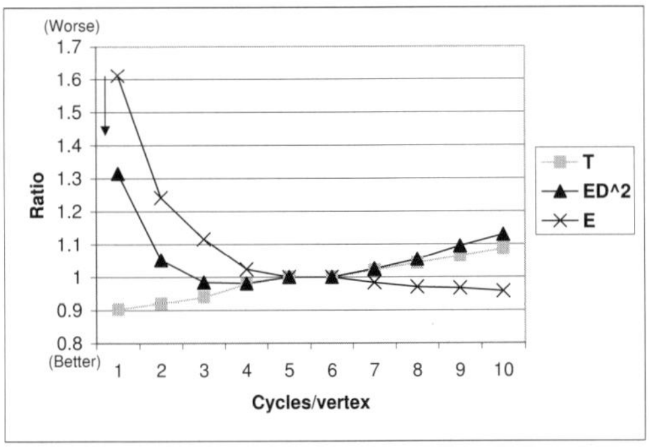

Figure 4: *Performance (T) and energy-efficiency data (ED^2 and E) for different vertex-processing rates. All results are normalized to our base case, which is 18 cycles/vertex.*

and performance. When considering the energy efficiency of different design choices, it is insufficient to simply look at average power, because $E = Pt$, and so if reducing average power comes at a cost of increased execution time, *more* energy may actually be consumed. Energy is also a poor metric when performance matters, since a reduction in energy only means that more power was saved than performance was lost. To cope with this, $E \cdot t$ has been proposed as a heuristic figure of merit, where the execution time t is also referred to as delay (D), hence the term "energy-delay product". This metric simply assumes that energy and performance are equally important. Better yet is the ED^2 product, which is more rigorous because it provides a metric of energy efficiency that is independent of the nominal supply voltage. It is therefore an accurate indicator of the unique contribution of microarchitectural or runtime power management that could not be achieved by simply choosing a different supply voltage at runtime. As an example, consider the the data above: if higher throughput is not needed—that is, the performance need not be any better than 1.0—then 4 cycles/tile is still the best configuration, because the chip's voltage can be reduced by 12%, yielding a 23% energy savings. The results presented here are only an illustration of the kind of tradeoff analysis that an infrastructure like Qsilver permits.

5.2.2. Multiple clock domains (MCD)

The notion of using multiple independent "clock domains" has been broached in both the GPU and CPU communities, e.g. [8, 14, 15] as a way to enhance clock speed by reducing the impact of clock skew and as a way to save power by allowing voltage and frequency to be dynamically reduced in clock domains that do not need to run at peak speed (dynamically changing voltage and frequency is often referred to as dynamic voltage scaling or DVS). In this paper, we focus on the power implications. Although MCD can provide substantial benefits, there are manufacturing and potential performance costs. MCD requires the ability to manufacture chips with independent voltage islands and independent clock domains, increasing testing costs; furthermore, each time the voltage and frequency settings are changed, the chip must stall while the voltage changes and the clock's phase-locked loop resynchronizes. A typical overhead for this is $10\,\mu s$.

In our GPU model, there are two natural clock domains—the vertex and fragment engines—with the fragment queue serving to decouple the two (see Figure 1). Since the GPU tends to alternate between being fragment and vertex bound, one or the other domain is typically stalled and can run at significantly lower voltage and frequency with minimal impact on overall fill rate. When the queue is full, this means the GPU is fragment bound, the vertex engine is effectively stalled, and it should reduce its voltage and frequency. Likewise, when the queue is empty, the GPU is vertex bound, the fragment engine is stalled, and it should reduce its voltage and frequency.

Specifically, we implemented a simple state machine that monitors the fragment queue. To avoid unnecessary changes in DVS setting and the associated overhead, the controller employs hysteresis by requiring the queue to stay within 10% of empty or full for 50,000 cycles before lowering the DVS setting by 10%. Then once the queue passes an appropriate high or low water mark to indicate that execution behavior is changing, the DVS setting is restored to its full value.

Figure 5 compares energy and performance both with and without MCD for three different leakage ratios. Again we use ED^2 as the figure of merit to trade off performance and energy and determine optimal energy efficiency. In fact, regardless of whether E or ED^2 is used, MCD is clearly energy efficient. For a 10% leakage ratio, MCD improves ED^2 by 11% with only a 1.5% performance loss. It achieves this by minimizing power dissipation when one or the other domain is stalled. Because leakage is exponentially dependent on voltage, MCD becomes even more advantageous at future technology nodes where leakage is a greater fraction of total power dissipation. When a domain is stalled but in a low voltage and frequency state, not only is dynamic power reduced, but leakage power is dramatically reduced as well. With the leakage ratio at 50%, the ED^2 improvement reaches 30%. These represent energy savings of 13–30%.

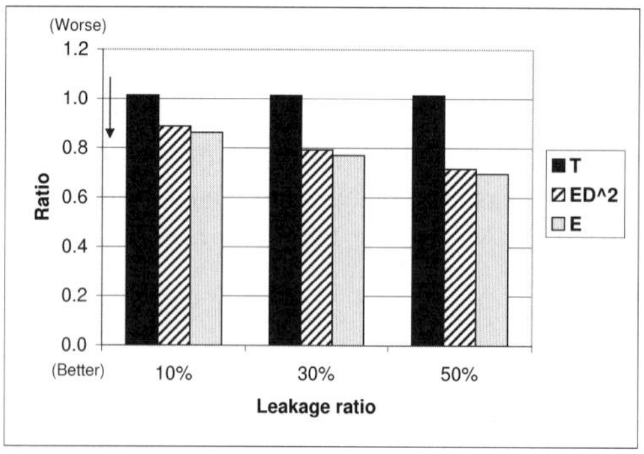

Figure 5: *Performance and energy-efficiency data of MCD for different leakage ratios. All results are normalized to the base case (with no MCD) for the appropriate leakage ratio. Smaller y-axis values are better, representing better performance and better energy efficiency.*

6. Discussion and limitations

Some tradeoffs accompany the decision to use Chromium, rather than writing or modifying a software rendering system such as Mesa [12], to generate the instrumented trace. Chromium's plug-in architecture makes it fast and simple to further annotate the stream by writing additional SPUs, or adding functionality to existing SPUs. We feel that the framework is quite flexible and, with some ingenuity, can incorporate a wide range of architectural experiments. For example, we plan in future work to explore the effects of Z_{min} culling [2] and Z_{max} or *hierarchical Z-buffer* culling [5, 11]. Typical hardware implementations of these ideas organize the depth buffer into 8×8 pixel tiles, and track the maximum or minimum depth, respectively, within the tile. If a given triangle is known to lie further than the maximum depth of a tile (for example, the distance to the triangle's nearest vertex is greater than that maximum depth), rasterization may be skipped in that tile; if the triangle's furthest vertex is closer than the tile's minimum depth, rasterization may proceed without requiring any Z-read operations.

To incorporate such tiled Z-buffer schemes, we propose to add another scratch buffer, with $1/8^{th}$ the linear resolution of the frame buffer, to the Qsilver annotator. This buffer stores at each pixel the maximum and minimum depth of the corresponding 8×8 tile of the framebuffer, and is updated after every primitive by running a min-max fragment program on the pixels of the scratch buffer affected by that primitive (easily accomplished by rasterizing a "fat" version of the primitive, or its bounding box, into the low-resolution scratch buffer). The min-max fragment program simply pools the pixels in the corresponding tile of the full-resolution depth buffer (bound as an input texture) and computes their min and max values. Given this low-resolution buffer, we can easily use an occlusion query and a special-purpose fragment program to render the primitive (at full resolution) and count how many fragments would pass the Z_{min} or Z_{max} tests.

Another motivation for using a hardware-accelerated annotation process via Chromium is speed: the use of hardware rasterization can be considerably faster than using an instrumented software renderer, typically in scenarios (high resolutions, full-screen stencil shadow buffer effects, etc) where the system is heavily fill-bound. Indeed, on our test platform a typical 50-frame trace from Enemy Territories runs $1.8\times$ faster through the Qsilver annotator than through uninstrumented Mesa at 1280x1024, and $2.6\times$ faster at 1600x1200.

It must be noted, however, that incorporating the annotations necessary to evaluate new architectural embellishments can be expensive, since every additional rendering pass and scratch buffer increases the number of pipeline flushes, context switches, and rendering calls. Furthermore, there are fundamental limitations to what data can easily be collected by our hardware-based annotator using fragment programs and occlusion queries. Our annotation approach is well-suited for collecting aggregate information: how many fragments get rasterized, pass the depth test, get textured with magnification versus minification, etc. While such information is sufficient for a wide range of studies, some experiments may require precise information about the positions or values of individual fragments. For example, it might prove difficult to study the effects of a Z-compression scheme in which memory bandwidth depended on the exact contents of the Z-buffer. In such a situation a user would likely wish to revert to instrumenting a software rasterizer, such as in Mesa, to track and collect the necessary data. Conveniently, Mesa implements the OpenGL API and can be incorporated directly into Qsilver's Chromium framework. This would allow the user to use the rest of the Qsilver framework and to re-use existing OpenGL traces, etc. It may even be possible to mix Mesa and hardware-accelerated OpenGL using different SPUs in Qsilver, for example running the slower Mesa rasterizer only for polygons or rendering intervals for which a particular architectural embellishment (e.g. a texture caching or compression scheme) was being evaluated.

7. Conclusions and future work

We have described Qsilver, a framework for power and performance analysis of graphics hardware. We have used Qsilver to develop a prototype power and performance model of a hypothetical GPU, and shown how this model allows:

- Fine-grained analysis of rendering behavior and bottlenecks, which can be used to optimize rendering systems.
- Architectural power and performance tuning, such as balancing the fragment-engine throughput to match the vertex engine's throughput.
- Evaluation of new architectural techniques for performance or power, such as Z_{min} culling or the use of DVS across multiple clock domains.

Our goal in developing Qsilver is to create a tool that will not only serve as a useful system for application performance analysis, but also stimulate research on GPU architecture and power-aware design in the same way that similar capabilities in the CPU community have fostered an explosion of research. It is important to note that we have designed Qsilver to be sufficiently flexible to model a wide array of architectures. We specifically wanted to ensure that it would be useful for modeling radical architectures to explore possible GPU architectures for future generations of graphics cards.

The next step in our work is refining our prototype to faithfully model a cutting-edge, programmable GPU. Then Qsilver makes possible a variety of interesting avenues for future work. For example, we would like to experiment with architectural innovations such as Delay Streams [1]. We are also particularly interested in modeling thermal effects. With a faithful performance and power model, and the addition of a more detailed floorplan, Qsilver can be augmented with a thermal model such as HotSpot [19] to dynamically simulate on-chip temperatures. This is an increasingly important area of research as high-performance GPUs become prohibitively costly to cool. For example, the tendency to oscillate between being vertex-bound and fragment-bound suggests that the GPU will in fact oscillate between thermal hotspots in these two regions. By smoothing out these hotspots spatially and temporally, temperature-aware design at the architectural level can significantly reduce cooling costs [16].

References

[1] T. Aila, V. Miettinen, and P. Nordlund. Delay streams for graphics hardware. *ACM Transactions on Graphics*, 22(3):792–800, 2003. 9

[2] T. Akenine-Möller and J. Ström. Graphics for the masses: a hardware rasterization architecture for mobile phones. *ACM Transactions on Graphics*, 22(3):801–808, 2003. 8

[3] T. Austin, E. Larson, and D. Ernst. Simplescalar: An infrastructure for computer system modeling. *IEEE Computer*, 35(4):59–67, Feb. 2002. 2

[4] D. Brooks, V. Tiwari, and M. Martonosi. Wattch: A framework for architectural-level power analysis and optimizations. In *Proceedings of the 27th Annual International Symposium on Computer Architecture*, pages 83–94, June 2000. 2

[5] N. Greene, M. Kass, and G. Miller. Hierarchical z-buffer visibility. In *Proceedings of the 20th annual conference on Computer graphics and interactive techniques*, pages 231–238. ACM Press, 1993. 8

[6] C. J. Hughes, V. S. Pai, P. Ranganathan, and S. V. Adve. Rsim: Simulating shared-memory multiprocessors with ilp processors. *IEEE Computer*, 35(4):40–49, Feb. 2002. 2

[7] G. Humphreys, M. Houston, R. Ng, S. Ahern, R. Frank, P. Kirchner, and J. T. Klosowski. Chromium: A stream processing framework for interactive graphics on clusters of workstations. *ACM Transactions on Graphics*, 21(3):693–702, July 2002. 1, 2

[8] A. Iyer and D. Marculescu. Power and performance evaluation of globally asynchronous locally synchronous processors. In *Proceedings of the 29th Annual International Symposium on Computer Architecture*, pages 158–68, May 2002. 7

[9] Y. Li, K. Skadron, Z. Hu, and D. Brooks. Understanding the energy efficiency of simultaneous multithreading. In *Proceedings of the 2004 International Symposium on Low Power Electronics and Design*, Aug. 2004. To appear. 2

[10] P. S. Magnusson et al. Simics: A full system simulation platform. *IEEE Computer*, 35(4):50–58, Feb. 2002. 2

[11] S. Morein. ATI radeon HyperZ technology. Presentation at Workshop on Graphics Hardware, Hot3D Proceedings, ACM SIGGRAPH/Eurographics, 2000. 8

[12] B. Paul et al. The mesa 3-d graphics library, 1993–2004. http://www.mesa3d.org/. 8

[13] M. Rosenblum, S. A. Herrod, E. Witchel, and A. Gupta. Complete computer simulation: The SimOS approach. *IEEE Parallel and Distributed Technology: Systems and Applications*, 3(4):34–43, Winter 1995. 2

[14] G. Semeraro, D. H. Albonesi, S. G. Dropsho, G. Magklis, S. Dwarkadas, and M. L. Scott. Dynamic frequency and voltage control for a multiple clock domain microarchitecture. In *Proceedings of the 35th Annual IEEE/ACM International Symposium on Microarchitecture*, pages 356–67, Nov. 2002. 7

[15] G. Semeraro, G. Magklis, R. Balasubramonian, D. H. Albonesi, S. Dwarkadas, and M. L. Scott. Energy-efficient processor design using multiple clock domains with dynamic voltage and frequency scaling. In *Proceedings of the Eighth International Symposium on High-Performance Computer Architecture*, pages 29–40, Feb. 2002. 7

[16] J. W. Sheaffer, K. Skadron, and D. P. Luebke. Temperature-aware GPU design. Poster at ACM SIGGRAPH, Aug 2004. 9

[17] A. L. Shimpi. NVIDIA GeForce4 - NV17 and NV25 come to life, Feb. 2002. http://www.anandtech.com/printarticle.html?i=1583. 6

[18] SIA. *International Technology Roadmap for Semiconductors*, 2001. 5

[19] K. Skadron, M. R. Stan, W. Huang, S. Velusamy, K. Sankaranarayanan, and D. Tarjan. Temperature-aware microarchitecture. In *Proceedings of the 30th Annual International Symposium on Computer Architecture*, pages 2–13, Apr. 2003. 9

[20] V. Zyuban. Unified architecture level energy-efficiency metric. In *Proceedings of the 2002 Great Lakes VLSI Symposium*, pages 24–29, Apr. 2002. 6

Figure 6: *The evolution of a frame in the* Enemy Territories *game.*

Graphics Hardware (2004)
T. Akenine-Möller, M. McCool (Editors)

Realtime Ray Tracing of Dynamic Scenes on an FPGA Chip

Jörg Schmittler, Sven Woop, Daniel Wagner[†], Wolfgang J. Paul, and Philipp Slusallek[‡]

Computer Science, Saarland University, Germany

Several example scenes rendered in realtime with 20 to 60 frames per second on the SaarCOR prototype hardware. From left to right: UT2003, Quake3, Conference and Terrain. Adding support for dynamically modifying these scenes actually reduces the hardware requirements.

Abstract

Realtime ray tracing has recently established itself as a possible alternative to the current rasterization approach for interactive 3D graphics. However, the performance of existing software implementations is still severely limited by today's CPUs, requiring many CPUs for achieving realtime performance.

In this paper we present a prototype implementation of the full ray tracing pipeline on a single FPGA chip. Running at only 90 MHz it achieves realtime frame rates of 20 to 60 frames per second over a wide range of 3D scenes and includes support for texturing, multiple light sources, and multiple levels of reflection or transparency. A particular interesting feature of the design is the re-use of the transformation unit necessary for supporting dynamic scenes also for other tasks, including efficient ray-triangle intersection as well as shading computations. Despite the additional support for dynamic scenes this approach reduces the overall hardware cost by 68 %.

We evaluate the design and its implementation across a wide set of example scenes and demonstrate the benefits of dedicated realtime ray tracing hardware.

Categories and Subject Descriptors (according to ACM CCS): I.3.1 [Hardware Architecture]: Graphics processors, Parallel processing I.3.7 [Three-Dimensional Graphics and Realism]: Ray Tracing, Animation

1. Introduction

Over the last two decades rasterization has become the dominant approach for interactive 3D graphics. However, due to its inability to directly access more than a single triangle at a time, rasterization is increasingly limiting the advancement of interactive 3D graphics and content creation. Without access to the entire scene even such apparently simple things as shadows, reflections, and indirect lighting are difficult to implement correctly and require non-trivial support from an application.

Application programmers must spend an increasing amount of their time working around these limitations instead of concentrating on their main job. Similarly, content creators struggle setting up special tricks and fakes necessary for at least approximating the required effects. As an example, rendering reflections of close-by objects requires to dynamically generate reflection maps depending on the current view and scene configuration. If applied to large or curved surfaces these maps must be applied carefully in order to avoid exposing the inherent inaccuracies of this approach.

† {schmittler,woop,cebron}@graphics.cs.uni-sb.de
‡ {wjp,slusallek}@cs.uni-sb.de

Ray tracing does not suffer from the same issues because it can directly, efficiently, and accurately simulate the path of light through an entire scene. While this allows highly realistic rendering and much simpler content creation, the required computations have been hugely expensive in the past and only allowed for off-line rendering. Recently, however, improved ray tracing algorithms and optimized software implementations already achieved realtime performance even on a single PC for simple scenes [Wal04]. But realtime performance in larger scenes or with non-trivial lighting effects still requires the combined power of many processors [Wal04].

Realtime ray tracing can become a viable alternative graphics technology if it can be offered within a single desktop computer. This requires highly parallel hardware that delivers the necessary floating point performance. In the past a number of proposals have been made, including the use of GPUs [PBMH02] and custom hardware [Gre91, GH96, MKS98]. However, none of these proposals was able to deliver realtime performance for non-trivial scenes while supporting all the usual ray tracing features.

In this paper we present an implementation of a fully featured ray tracing pipeline in hardware. Using a single FPGA chip running at 90 MHz it offers realtime rendering performance of 20 to 60 frames per second while supporting all important ray tracing features, like Phong-like shading, textures, reflections, transparencies, shadows, and indirect lighting even in non-trivial scenes.

The hardware also supports interactively changing the scene through an object-based approach that allows to affinely transform entire groups of triangles. The same approach can also be used to instantiate these objects multiple times at arbitrary locations in the scene.

This technique requires frequent transformations of rays between different coordinate spaces using a dedicated transformation unit. Unfortunately, this unit requires significant hardware resources while remaining idle for long periods of time during which other units perform different ray tracing computations.

Consequently, we use special algorithms for ray-triangle intersection computations as well as for important parts of the shading computations that re-use the available but idle transformation unit. This strategy greatly simplifies the intersection and shading units and reduces the overall hardware cost by up to 68 %.

1.1. Previous Work

Ray tracing offers many benefits in the context of a hardware implementation. In particular, it scales trivially and can easily be parallelized. For a detailed overview of the state-of-the-art in interactive ray tracing see [WPS*03].

Several approaches have already been realized on MIMD and SIMD architectures [GP89, GP90, LS91]. Exploiting this scalability by massive parallelization has recently allowed interactive ray tracing to be achieved in software. It was first realized on supercomputers [Muu95, KH95, PSL*99] and more recently, interactive performance has been brought to clusters of standard PCs [WBWS01, WPS*03].

The availability of interactive ray tracing required efficient algorithms to support dynamic scenes. Thus [RSH00] covered dynamic updates to scene data structures and [LAM01] investigated lazy evaluation for scene data. More recently [WBS03] presented dynamic scene management on a distributed rendering cluster. Furthermore there is the BART suite [LAM00], a well designed benchmark suite to evaluate implementations of dynamic ray tracing systems.

Our hardware implementation is based on an extended version of the software approach outlined in [WBWS01, WBS03] which has been shown to efficiently handle the BART suite. Section 5 shows that the prototype's performance is several times faster than the software version on a single CPU over a wide range of different dynamic scenes. Thus although we did not implement the BART suite for the prototype we expect it to run well on our hardware.

Besides using existing architectures, several special purpose hardware architectures for ray tracing have been developed. Initially hardware support was provided only for the intersection operation [Gre91]. Later DSPs were used to build PC-card based ray tracing accelerators [GH96]. Several volume ray casters on PC-cards have been developed [MKS98, PHK*99, Pfi01] and there is a commercially available hardware architecture for high quality off-line ray tracing [Hal01].

While these projects achieved remarkable results and work well as accelerators for ray tracing, none of them is capable of delivering full-screen realtime frame rates comparable to those of current rasterization hardware. Recently, a proposal for a complete hardware architecture for ray tracing has been simulated that supports static scenes [KiSSO02].

Instead of designing special purpose hardware for ray tracing, another interesting approach maps it to more general future architectures: Using a multi-processor system on a single chip [MPJ*00], Purcell has developed a highly optimized distributed ray tracer that would be capable of delivering interactive performance [Pur01]. More recently, ray tracing was also mapped to rasterization hardware using their programmable pipelines [PBMH02]. These two projects show that ray tracing can in principle be realized on such multi-processor single chip hardware architectures.

1.2. Overview of the Paper

In this paper we explore the design of special purpose hardware for ray tracing. In particular, this allows us to explore the performance and efficiency with which ray tracing can be implemented in hardware. As we will see later, the hardware efficiency of those implementions on GPUs and CPUs is rather low compared to our custom hardware implementation.

The design for our ray tracing hardware is based on the SaarCOR [SWS02] architecture and adapts it for an implementation on modern FPGAs. We discuss the necessary changes to the architecture and evaluate the implementation and its performance using example scenes from several different application scenarios.

This paper is structured as follows: We start with the presentation of a ray tracing algorithm for dynamic scenes which uses transformations and discuss the re-use of these transformations for other tasks. The architecture of our design and its prototype implementation is then described in the Sections 3 and 4. We analyze and evaluate the design in Section 5 before discussing possible consequences and future work in Section 6.

2. Ray Tracing of Dynamic Scenes

The basic ray tracing algorithm intersects rays with triangles and stores the intersection that is closest to the ray's origin. To increase performance, a *spatial index* is used that subdivides space and allows to quickly locate objects in space. Rays then only have to be intersected with objects located in those spatial regions that are pierced by the ray. With hierarchical indices the expected runtime of ray tracing becomes logarithmically in the scene complexity. This compares favorably to the linear complexity of a rasterization unit (unless the application implements similar techniques on top of the rasterization engine, i.e. using occlusion queries).

However, the creation of the index is at least linear in the number of scene primitives, which seems to remove the advantage for ray tracing in the case of dynamic scenes. Fortunately, often large parts of dynamic scenes remain static for long periods of time allowing for amortizing the cost of building a spatial index over many frames. Additionally, we can trade off the building cost against the quality of an index depending on its expected live time allowing us to spend less time for building an index that cannot be amortized over many frames.

We base our approach on [WBS03] where primitives are split into separate *objects* that are either completely replaced by modified content or which move under a single affine transformation with respect to the scene. We focus on the second approach which allows us to pre-compute and re-use a separate *bottom-level spatial index* for each such object. Each such index can then be positioned in the scene with an affine transformation and is inserted into a *top-level spatial index*, which must be rebuild whenever any object moves.

When a ray encounters an object in the top-level index it is transformed into the object's coordinate space and can now continue traversing the pre-computed index of the object [WBS03]. Objects can be instanciated multiple times simply by storing multiple transformations together with references to the original object. This approach maintains the logarithmic cost in scene complexity even for dynamic scenes with instances.

The same approach can also be applied to a hardware implementation of ray tracing but this requires that a functional unit is provided for performing the transformation of the rays. This unit is expensive in the number of floating-point units that are required but remains idle most of the time. Consequently, we investigated other uses of the transformation unit in ray tracing, which are discussed below.

2.1. Ray Triangle Intersection

As the core part of ray tracing is the computation of the intersection between a ray and a triangle, this operation has received much attention in research (see [MT97, Eri97, Wal04]). In this section, we present a slightly extended version of Arenberg's algorithm [Are88] that is optimal for our purposes as it utilizes the transformation unit to perform most of the ray-triangle intersection calculations and also computes the dot product between ray direction and the normal of the triangle for free.

Figure 1 illustrates the *unit triangle intersection method* which consists of two stages: First the ray is transformed using a triangle specific *affine triangle transformation* to a coordinate system in which the triangle is the *unit triangle* Δ_{unit} with the vertices $(1,0,0), (0,1,0)$, and $(0,0,0)$. In the second stage, a much simplified intersection test of the transformed ray with the unit triangle is performed.

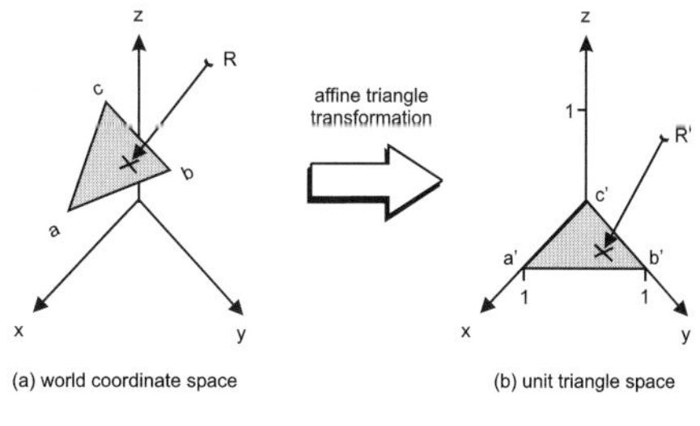

Figure 1: *The unit triangle intersection method consists of two stages: First the ray is transformed, using a triangle specific affine triangle transformation. In the second stage, a simple intersection test of the transformed ray with the unit triangle is performed.*

2.1.1. Affine Triangle Transformation

The affine triangle transformation to a triangle $\Delta = (A,B,C)$ with $A,B,C \in \Re^3$ is an affine transformation $T_\Delta(X) = m \cdot X + N$ with $m \in \Re^{3,3}$ and $X,N \in \Re^3$ that maps the triangle Δ to the unit triangle Δ_{unit} such that the normalized normal $N = \frac{(A-C)\times(B-C)}{|(A-C)\times(B-C)|}$ of the triangle is mapped to the normal $N_{unit} = (0,0,1)$ of the unit triangle.

The inverse T_Δ^{-1} of T_Δ can easily be described by the following equations:

$$T_\Delta^{-1}\begin{pmatrix}1\\0\\0\end{pmatrix} = A \quad T_\Delta^{-1}\begin{pmatrix}0\\1\\0\end{pmatrix} = B$$

$$T_\Delta^{-1}\begin{pmatrix}0\\0\\0\end{pmatrix} = C \quad T_\Delta^{-1}\begin{pmatrix}0\\0\\1\end{pmatrix} = N$$

These equations map the vertices of the unit triangle to the vertices of the triangle Δ and the normal N_{unit} to N. The solution for T_Δ^{-1} takes the form:

$$T_\Delta^{-1}(X) = \begin{pmatrix} A_x - C_x & B_x - C_x & N_x - C_x \\ A_y - C_y & B_y - C_y & N_y - C_y \\ A_z - C_z & B_z - C_z & N_z - C_z \end{pmatrix} \cdot X + \begin{pmatrix} C_x \\ C_y \\ C_z \end{pmatrix}$$

The transformation T_Δ^{-1} is unique and well defined. If the triangle is not degenerate its inverse T_Δ exists and is a one-to-one affine transformation.

2.1.2. Unit Triangle Intersection

A ray $R = (O,D)$ with the origin $O \in \Re^3$ and direction $D \in \Re^3$ is intersected with a triangle Δ by transforming R using T_Δ to the *unit triangle space* and intersecting the transformed ray with the unit triangle. We do not directly compute the point of intersection P but the *intersection parameter* $t \in \Re$, such that $P = O + t \cdot D$. The parameter t and the barycentric coordinates of P within Δ do not change under a one-to-one affine transformation. Thus it is equivalent to compute the ray-triangle intersection in world coordinate space or in unit triangle space.

This transformation greatly simplifies the intersection computation of the ray with the triangle. Let $R' = T_\Delta(R) = T_\Delta(O,D) = (m \cdot O + N, m \cdot D) = (O',D')$ be the ray transformed to the unit triangle space, then the intersection can be computed by:

$$t = -\frac{O'_z}{D'_z}, \quad u = O'_x + t \cdot D'_x, \text{ and } v = O'_y + t \cdot D'_y.$$

2.1.3. Dot Product Preservation

An additional analysis of Arenberg's algorithm [Are88] allows to compute the dot product between the ray direction D and the normal of the triangle for free. Since the normalized normal of the triangle was mapped to the normal N_{unit} of the unit triangle, the dot product is simply $D' \cdot N_{unit} = D'_z$. This property can be exploited if designing shading units, as shaders typically require the cosine between ray direction and geometry normal for color calculation and ray generation (see next section).

2.1.4. Further Applications

This concept of first transforming a ray to a generic coordinate space before intersecting it can also be applied to many other types of geometric primitives, e.g. boxes, ellipsoids, cylinders, etc. The advantage is that only a single representation (the transformation) needs to be stored together with a flag indicating the type of primitive. Additionally only a much simpler and smaller primitive-specific second stage intersection unit must be added.

2.2. Shading

Shading a ray consists of two parts: computing the local scattering of light and spawning new rays to gather the incident light from certain directions. The transformation unit is well suited for the second part as discussed below.

2.2.1. Transformation of the Normal

For advanced shading effects such as reflection and refraction, the normal of the triangle is used to calculate secondary rays. This normal is specified in object coordinate space, but needs to be transformed to world coordinate space. Obviously this transformation can be performed using the transformation unit.

2.2.2. Ray Generation

Spawning of a ray requires several floating-point operations. These operations can be specified using a transformation $T(X) = (\,A\;B\;C\,) \cdot X + D$. For simplicity reasons we write $T = [A,B,C;D]$ to specify a transformation.

Using transformations for ray generation allows for using of the transformation unit and thus reduces the complexity of the shading unit. Therefore we compute a new ray R_{new} by providing a transformation T_{new} and an initial ray R'_{new} to the transformation unit which then calculates R_{new} as input to the remaining ray tracing units.

Primary Rays We specify a camera by its position C_p and an orthonormal basis $\{C_r, C_u, C_d\}$ formed by the right-vector, the up-vector, and the viewing direction. The values $x,y \in [-1,\ldots,1]$ parameterize the screen space with a unit view frustum of 45 degree. For each pixel (x,y) on the screen the initial ray $R'_{init} = ((0,0,0),(x,y,1))$ is mapped using the transformation $T_{init} = [C_r, C_u, C_d; C_p]$ to the primary ray $R_{init} = (C_p,\, x \cdot C_r + y \cdot C_u + C_d)$.

Shadow Rays Shadow rays can be calculated very easily given the incident ray $R = (O,D)$, its intersection parameter t, and the position of the light source L. Using $T_{shadow} = [L,O,D;0]$ and $R'_{shadow} = ((1,0,0),(-1,1,t))$ yields the shadow ray $R_{shadow} = (L,\ O+t\cdot D-L)$.

Reflection Rays The calculation of the reflection ray requires the normal N of the triangle, the normalized ray direction D, the cosine c between N and D, and a small positive value e to avoid self intersections. The ray is computed using $T_R = [D,N,D;O]$ and the initial ray $R'_{refl} = ((-e,0,t),(1,-2\cdot c,0))$ yielding the reflection ray $R_{refl} = ((O+t\cdot D-e\cdot D),(D-2\cdot c\cdot N))$.

Transparency Rays For calculating a transparency ray simply the same transformation T_R as for the reflection ray can be used with the initial ray $R'_{transp} = ((e,0,t),(1,0,0))$. A transparency ray then evaluates to $R_{transp} = ((O+t\cdot D+e\cdot D),D)$.

Refraction Rays Unfortunately, the situation for refraction rays is a bit more complicated but some part of the nonlinear refraction calculation can be performed using the transformation T_R as in the reflection case. In a first step we compute $\mu = \eta\cdot c-\sqrt{1-\eta^2+(\eta\cdot c))^2}$ using $c = N\cdot D$, which has been computed by the triangle intersection for free. The initial ray $R'_{refr} = ((t,0,e),(-\eta,\mu,0))$ is then mapped by T_R to the refraction ray $R_{refr} = (O+t\cdot D+e\cdot D,\ \mu\cdot N-\eta\cdot D)$ of a surface with the index of refraction η.

2.3. Implementation Issues

The transformation unit must perform a 3×3 matrix multiplication and a vector addition for the origin of the ray as well as another matrix multiplication for the ray's direction. Since a matrix multiplication is very costly in terms of hardware resources, only one matrix multiplication unit together with the vector addition is implemented. The transformation of the direction is performed by feeding zeros to the correct coefficients. This requires two steps of the transformation unit to completely transform a ray. Fortunately, in many cases the transformation unit is used to transform a sequence of several rays sharing the same origin (e.g. primary rays or shadow rays). In those cases this allows for reducing the workload of the transformation unit for a sequence of n rays from $2n$ to only $n+1$ steps.

In order to evaluate the hardware savings, we compare the number of floating point units of four different ray tracer designs. The first variant RT_{static} consists of a static ray tracing pipeline using a ray-triangle intersection algorithm I_w based on Wald [Wal04] and simple shading unit S that only performs the geometric calculations on rays with floating point operations. The second variant RT_{Dyn1} is a dynamic ray tracer using the optimizations described above and therefore only contains floating point units in the transformation unit T and in the small primitive intersection unit. Variant RT_{Dyn2} is a dynamic ray tracer build with standard methods using the units T, S and I_w. Similar to this variant, RT_{Dyn3} only differs in the ray-triangle intersection algorithm I_p based on Plücker [Eri97], which is a bit cheaper than Moeller-Trumbore's [MT97]. All designs have the traversal unit K in common.

Table 1 shows that with the optimizations presented in this paper, a dynamic ray tracing chip is already significantly cheaper in terms of floating point units than an optimized static ray tracer even though it offers additional new functionality. Furthermore, re-using the transformation unit for other purposes cuts the number of floating point units at least in half.

FP cost ratio	without Traversal	full design
RT_{Dyn1} / RT_{static}	66%	75%
RT_{Dyn1} / RT_{Dyn2}	46%	57%
RT_{Dyn1} / RT_{Dyn3}	23%	32%

Table 1: *Comparisons of the cost of four different ray tracers measured in floating point units. It shows that the optimizations presented in this paper allow to significantly reduce the hardware costs.*

3. SaarCOR Hardware Architecture

We build on the SaarCOR ray tracing architecture first presented in [SWS02], which we extend with support for dynamically moving objects and multiple instantiations of objects. In [SWS02] the focus was on the ray tracing core of a static ray tracer. It was shown, that the SaarCOR architecture scales well in the number of ray tracing pipelines, in particular due to its good caching behavior and consequently low memory bandwidth requirements. With approximately the same amount of hardware it achieved roughly the same performance compared to a rasterization based graphics card.

In order to remove the need to have the entire scene in graphics memory, [SLS03] presented a virtual memory architecture for ray tracing. This fully transparent technique uses graphics memory only as a cache and loads necessary scene data from host memory on-demand. Over a wide variety of scenes even a slow standard PCI bus is able to support a ray tracer with only a small impact on the rendering performance.

3.1. Dynamic SaarCOR Architecture

The SaarCOR architecture for dynamic ray tracing (see Figure 2) is designed for a custom chip that is connected to the host system bus and to several external memory chips, all placed on a single PC board. Camera settings and scene data are uploaded by the host. The memory chips store the scene

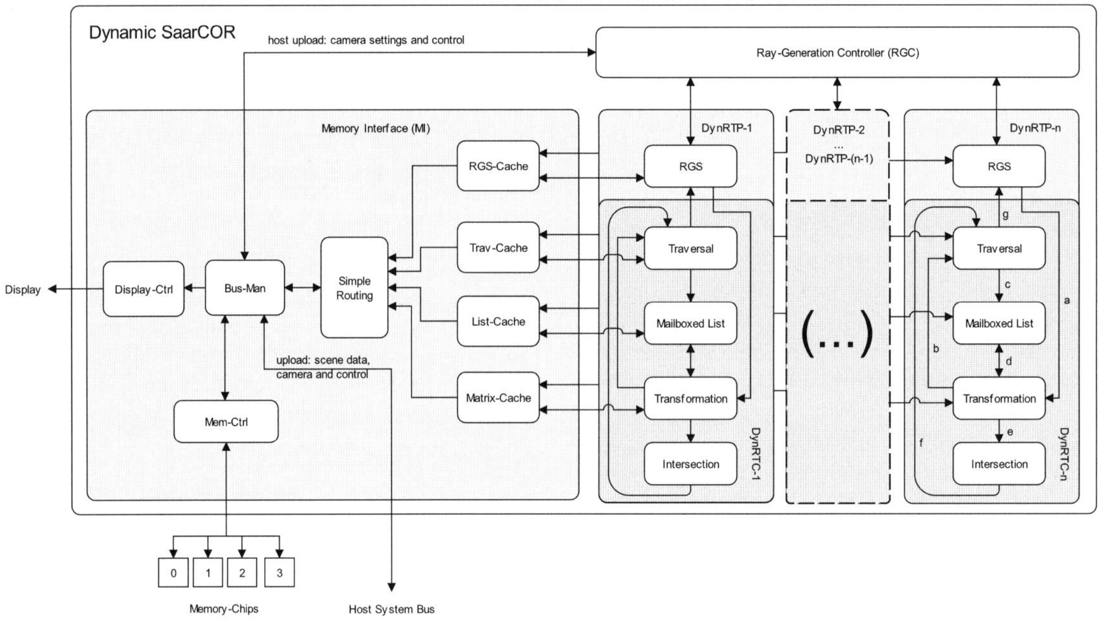

Figure 2: *The SaarCOR architecture for dynamic ray tracing is built into a custom chip connected to external I/O and memory. The chip is split into three functional sections: The dynamic ray tracing pipeline (DynRTP) which contains the dynamic ray tracing core (DynRTC) and the ray generation and shading unit (RGS). The RGS-units are controlled by a ray generation controller (RGC). The memory interface (MI) manages the connection between the rendering units, the memory, and the host system bus. Scalability is achieved by supporting several DynRTPs in parallel. Please note the simple routing scheme that uses only point-to-point connections.*

data including geometry, top- and bottom-level spatial index structures (in our case KD-trees), materials and shaders, as well as the frame buffer. The rendered image is displayed by the *Display-Controller*.

The core ray tracing algorithm is contained in the dynamic ray tracing pipeline (DynRTP). The tracing of a ray is started at the ray generation and shading unit (RGS) which generates an initial ray R' and transformation T which are sent to the dynamic ray tracing core (DynRTC) via path a of Figure 2. The transformation T is applied to the ray R' by the transformation unit and the transformed ray is sent to the traversal unit via path b.

The traversal unit then starts traversing the ray through the top-level KD-tree until a leaf node is found. The ray with the list of objects is then forwarded via path c to the mailboxed list unit, which sequentially fetches objects referenced in the list while omitting objects that have already been visited by the same ray. This technique greatly improved performance by avoiding the cost of multiple traversals of the same object.

Path d is used to send the ray together with the address of a transformation to the transformation unit which maps the ray into object space. The ray is sent to the traversal unit via path b to start the bottom-level traversal.

The same process starts over except that leaf nodes now contain lists of triangle transformations that are used to transform the ray into triangle space before being forwarded to the intersection unit via path e. The intersection results are returned to the traversal unit via path f and the operation continues at the list unit until all elements have been processed (similarly at the object level). Finally, results are handed back to the RGS unit for shading via path g.

In order to hide memory and other latencies multiple rays are being processed simultaneously using a multi-threading approach. Optionally packets of rays are used to further reduce the memory bandwidth in the ray tracing core. Scheduling of rays in this complex pipeline is trivial requiring only minimal state memory and simple logic to route rays to the correct next stage.

All memory accesses of the DynRTP are handled by the memory interface (MI) which contains individual caches for each type of data (i.e. shading data, KD-tree nodes, lists, and matrices for dynamical objects and triangles). All caches share the same connection to the memory via the *Simple Routing* unit. This unit acts like a multiplexer for memory requests and uses a labeled broadcast for memory responses [SWS02].

The memory controller (Mem-Ctrl) manages accesses to external memory chips and uses a simple reordering algorithm that optimizes the effective bandwidth of the memory chips by avoiding unnecessary switching of memory-banks. This performs close to optimally because the memory access pattern of tracing many rays simultaneously are almost random and thus statistically well distributed across the banks, which allows for keeping them all busy.

For scalability several DynRTPs are supported. The work performed in each DynRTP is controlled by the ray generation controller (RGC) which schedules screen-space pixels. This performs dynamic load balancing between the ray tracing pipelines while requiring only minimal bandwidth between the RGC and each of the DynRTPs.

The architecture exploits coherence between adjacent rays by using packets of independent rays as described in [SWS02]. This significantly reduces the bandwidth to the caches as only one memory request is performed per packet while the fetched data can be used for all rays. This greatly simplifies the interface to the cache and increases the scalability of the architecture.

3.2. Shading

The SaarCOR architecture is designed to support different types of shading processors: multiple general purpose processors like standard RISC-CPUs or shading units similar to today's rasterization hardware. A detailed description of those features is beyond the scope of this paper.

In this paper we focus on a fixed function shading pipeline that was implemented in the prototype. This shader supports multiple light sources, multiple levels of reflection or transparency and Phong-like shading with bilinearly filtered textures. For each ray that is shaded, shading data is fetched from memory, including material color, normals, texture-coordinates, and the texture base address. Additional memory requests are done for texture lookup. Each material and texture color consists of RGB data and values specifying the degree of specularity and opacity.

Rays are calculated using the formulas presented in Section 2.2 by forwarding the transformation and initial ray to the transformation unit via path *a* in Figure 2. Since for simplicity all colors are currently processed as 24 bit integers no floating point operations are required in the shading units.

4. Prototype and Implementation

While the original SaarCOR architecture already provided promising simulation results, it was never proven in the real world. Consequently, we decided to develop a prototype that demonstrates the usefulness and performance of a complete hardware implementation of ray tracing. The prototype was developed on FPGA hardware in less than six month by a small dedicated team of 3 students. It already provides realtime frame rates and all important features of ray tracing. We describe and evaluate this prototype in the following sections.

4.1. Development Platform

The SaarCOR prototype is build using a Xilinx Virtex-II 6000-4 FPGA [Xil03], that is hosted on the Alpha Data ADM-XRC-II PCI-board [Alp03]. The board contains six independent banks of 32-bit wide SRAM (each 4MB) running at the FPGA clock speed, a PCI-bridge, and a general purpose I/O-channel. This channel is connected to a simple digital to analog converter implementing a standard VGA output supporting resolutions of up to 1024×768 at 60 Hz.

The SaarCOR design was specified using JHDL [Bri03], an OpenSource high-level hardware description language allowing for fast and painless prototyping with the flexibility of parameterized designs. Additionally, we used Xilinx tools for schematic entry and low level synthesis. The system was completely developed under Linux.

4.2. Implementation Issues

The floating-point units of SaarCOR implement single operations, such as addition, comparison, and multiplication. They were modeled in JHDL with a parameterized precision in order to evaluate its effect on performance and rendering quality. The results of several benchmark scenes showed that 24 bit floating-point numbers with 16 bit mantissa provide a good compromise between accuracy and hardware cost. This format is similar to those used in current graphics boards by ATI.

Two of the six banks of SRAM are used to implement a double-buffered frame buffer with resolutions of up to 1024×768 pixels. A third SRAM stores all shading data, while the remaining three banks contain the KD-tree and the object and triangle transformation matrices.

The memory bandwidth for shading is rather low because only a small amount of shading data is required (28 byte per ray with bilinear texture filtering), and only the final pixel color needs to be written to the frame buffer. No bandwidth is required as for read-modify-write cycles of a Z-buffer or the overhead of overdraw. The resulting bandwidth is low enough (see Section 5) that it did not even require caches (which nonetheless would be fairly effective as shown by our simulations).

In [SWS02] it has been shown that 4 times more traversal operations than intersection operations need to be computed during the rendering of typical scenes. Since a single FPGA can only hold one intersection unit (see Section 4.3) we use packets of four rays which are traversed in parallel and are intersected sequentially. If all units would be fully utilized, they would require a raw bandwidth of 2 GB/s when running at 90 MHz. Due to small direct mapped caches that already yield very good hit rates and non-perfect utilization we easily reduce the bandwidth to a small fraction of the 1 GB/s available to the DynRTC (see Section 5).

4.3. Hardware Complexity

Table 2 lists the hardware resources required by a single ray tracing pipeline measured in the number of floating-point units for addition, multiplication, division, and comparison. The rightmost column also provides the amount of internal memory used for instance for ray-data and stacks that store 64 KD-tree traversal states per packet. These numbers include all additional index structures of the caches and dual port memory bits are counted as 2 bits. It is obvious that the arithmetic complexity and the internal memory requirements are extremely low.

Unit	Add	Mul	Div	Cmp	Mem
Traversal	4	–	4	13	44.5 KB
Mailboxed List	–	–	–	–	0.8 KB
Transformation	9	9	–	–	9.3 KB
Intersection	3	2	1	–	–
DynRTC-Cache	–	–	–	–	15.6 KB
Shader	–	–	–	–	4.8 KB
Total	16	11	5	13	75.0 KB

Table 2: *Complexity of one ray tracing pipeline measured in floating-point units for addition, multiplication, division, and comparison, respectively. The rightmost column also lists the internal memory requirements including any metadata and global state, such as parameters for 8 lightsources. Each DynRTP uses 32 threads and contains caches that store 512 data-items each.*

Since the FPGA provided more capacities than required, we implemented a system with 64 threads, shading with support of 256 light sources, PCI interface, VGA interface, and performance counting infrastructure. Still our design only utilizes 56% of the FPGA's logic cells and 78% of the FPGA's memory resources including wasted resources due to memory layout and mapping constraints. The prototype runs at a frequency of 90 MHz and delivers a total of 4 billion FLOPs.

Much larger FPGAs are already available today that have about 60% more logic cells and four times more memory and multiplier blocks. This would support at least two additional ray tracing pipelines because the memory and multiplier resources have been the most limiting factor in the design of the prototype.

Our design also compares well to today's high end rasterization hardware. For instance, Nvidia's GeForce 5900FX [Nvi04] contains 125 million transistors (3-times more than Intel's Pentium-4). Its 400 FP-units running at a frequency of 500 MHz yield 200 billion FLOPs, which is 50-times the performance of the SaarCOR prototype. Despite that a direct comparison is not really possible, these resources would allow for quite a large number of parallel ray tracing pipelines.

As shown below the memory bandwidth of SaarCOR including *un-cached* shading is mostly far less than 300 MB/s plus additional 135 MB/s to display the image in 1024×768 at 60 Hz. As today's graphics boards offer more than 30 GB/s external memory bandwidth, this would support more than 100 independent ray tracing pipelines.

Together this indicates that SaarCOR could be scaled by one to two orders of magnitude going from our current FPGA technology to one used by today's rasterization engines.

5. Benchmark-Scenes and Results

The previous incarnations of SaarCOR [SWS02, SLS03] presented a ray tracer for static scenes and a virtual memory architecture, which were simulated and evaluated on the register transfer level using a cycle accurate simulator. We used the same simulator also for the Dynamic SaarCOR architecture as presented above. This simulation used a "standard" SaarCOR configuration with four parallel DynRTCs and provided essentially the same results as for the previous architectures: high cache hit rates, very low external memory bandwidth due to the use of larger packets and caching, good usage of all pipeline stages, high scalability, and lower hardware requirements for similar performance when compared to rasterization. Please refer to these papers for more detail.

In the following we focus on providing results of measurements of the FPGA prototype as discussed in Section 4. Even though the specific implementation and its parameters (such as memory latencies, etc.) are very different, these measurements confirm and verify the simulation results.

In particular we provide comparisons of the hardware prototype with the highly optimized software implementation of the OpenRT realtime ray tracing system [Wal04]. It uses SSE-optimized code on packets of four rays and algorithms almost identical to those of the prototype. The software runs on a Intel Pentium-4 with 2.66 GHz and 1 GB RAM. The SaarCOR prototype runs at 90 MHz on the configuration described above. The host computer's influence on the results are completely negligible since all computations are performed directly on the SaarCOR hardware except for uploading changes to the scene between frames. This uploading can be performed in parallel to rendering.

The configurations for the measurements differ in two ways from results presented in earlier papers: we use new algorithms for building KD-trees that greatly improve the quality and thus the traversal cost of KD-trees [Wal04]. This speedup is included in both the software and the hardware measurements given below. Furthermore all measurements include fully textured shading, which is quite costly on CPUs [Wal04].

5.1. Scenes used for Benchmarking

Table 3 lists the scenes and its parameters used to evaluate our prototype. Screen shots of some example scenes are show in the images on the first page of this paper. Full resolution images and videos can be found on the project's home page at `http://www.SaarCOR.de`.

To cover a wide range of 3D applications, we use a variety of scenes ranging from simple scenes, such as a simple room with table and chairs as in *Scene6* up to huge scenes with hundreds of millions of triangles using instantiation of complex objects as in *SunCOR* (5,622 sunflowers each consisting of 33,288 triangles). We render the full details of all scenes with no level-of-detail mechanisms, as ray tracing handles such complex scenes easily due to its logarithmic computational complexity and its output sensitive computation that only ever touches visible parts of the scene.

Other more realistic examples are taken from computer games, such as *Castle* [Act02], *Quake3* [Id-99], and *UT2003* [Epi03]. While *Quake3* and *UT2003* only contain static meshes, *Quake3-p* also includes 16 moving players and monsters. The *Castle* scene also shows some nice ray tracing effects including multiple reflections. The videos to this paper include a small shooter game running on top of the ray tracer using the *UT2003* environment.

The other three scenes, *Office*, *Conference*, and *Terrain* provide more examples of indoor and outdoor scenes and show an office, a conference room with many light sources, and a large outdoor scene with complex trees casting detailed shadows.

The plug-and play concept of ray tracing allows to quickly and easily create such benchmark scenes: every object is described by its own self-contained shader independent of any other objects or shaders. All global effects resulting from interactions of multiple objects and shaders are computed correctly and on-demand during ray tracing. No manual tweaking or preprocessing is required except for building the spatial index structures.

This property of ray tracing greatly simplifies and speeds up content creation compared to rasterization, which requires many tricks to obtain approximations of global effects and care must be taken to avoid exposing their limitations.

5.2. Results

Table 3 compares the performance of the SaarCOR prototype to the OpenRT software implementation when rendering the benchmark scenes at 512×384 pixel using primary rays only, but including fully textured shading. It shows, that although the CPU is clocked 30 times faster than the SaarCOR prototype, the hardware is still 3 to 5 times faster. Thus, the 90 MHz prototype is theoretically equivalent to an 8 to 12 GHz CPU.

Looking at the raw FLOPs of the underlying hardware and comparing the resulting frame rates shows that SaarCOR uses its floating point resources 7 to 8 times more efficiently than the SSE-optimized OpenRT software on a Pentium-4. This means that even the highly optimized software ray tracing code uses the available floating point hardware only to a small fraction, indicating that the current CPU designs are non-optimal for ray tracing (even ignoring their insufficient maximum floating point performance).

In comparison, the fastest published ray tracer on GPUs delievers 300K to 4M rays per second on an ATI Radeon 9700PRO [Pur04]. In contrast, our simple FPGA prototype already achieves 3M to 12M rays per second at a much lower clock rate and using only a fraction of both the floating point power and the bandwidth of this rasterization hardware. An implementation in a comparable ASIC technology should allow us to significantly scale the ray tracing performance even further by at least an order of magnitude.

			#frames per second		
Scene	#triangles	#objects	SaarCOR	OpenRT	Speed-Up
Scene6	806	1	60.8	12.9	4.7
Castle	20 891	8	23.8	9.2	2.6
Office	34 312	1	48.9	10.4	4.7
Quake3	39 424	1	33.6	11.1	3.0
Quake3-p	52 790	17	26.7	7.9	3.4
UT2003	52 479	1	25.4	8.0	3.2
Conference	282 805	54	22.1	8.1	2.7
Terrain	10 469 866	264	15.9	3.5	4.5
SunCOR	187 145 136	5622	32.1	7.5	4.3

Table 3: *This table lists our fully textured benchmark scenes with their complexity (number of triangles and dynamic objects). The three rightmost columns compare the performance of the SaarCOR prototype with only one rendering pipeline running at 90 MHz to the OpenRT software ray tracer with SSE-optimized code on an Intel Pentium-4 2.66 GHz. The images were rendered at* 512×384 *pixel using primary rays only but including fully textured shading. It shows, that although the CPU is clocked 30 times faster than the SaarCOR prototype, the hardware is still 3 to 5 times faster. While all scenes use bilinear filtered textures, Scene6 and Office were rendered with unfiltered textures since otherwise the shading bandwidth would slightly limit the performance (10% at most).*

Table 4 provides a more detailed view on the external memory bandwidth and the cache hit rates. These measurements were taken at a screen resolution of 1024×768. In addition this table lists the usage ratios for the units of the DynRTC.

		Usage Rate [in %]				Cache Hit Rate [in %]			Ext. Bandwidth [in MB/s]		
Scene	fps	Trav	List	Transf	Int	Trav	List	Tranf	DynRTC	Shading	Total
Scene6	15.3	68	16	85	41	99	97	99	8	218	226
Castle	5.9	73	18	74	48	99	82	94	50	138	188
Office	12.4	76	16	72	36	99	71	90	69	246	246
Quake3	8.5	87	15	45	21	99	49	79	104	197	301
Quake3-p	6.8	92	13	35	16	99	66	82	65	157	222
UT2003	6.5	82	19	66	42	98	63	86	105	152	257
Conference	5.7	89	25	51	28	98	63	78	135	132	267
Terrain	4.2	80	27	34	18	97	27	43	283	98	381
SunCOR	8.7	46	36	29	12	90	2	6	513	202	715

Table 4: *This table provides details on the performance of the SaarCOR prototype running at 90 MHz and with a resolution of* 1024×768 *pixels using primary rays only but with bilinear-filtered textured shading. We provide the usage for each unit of the DynRTC, the hit rates of the all caches, as well as the external memory bandwidth (excluding frame buffer readout for display). It shows that multi-threading allows to efficiently keep most of the units busy and high hit rates are achieved even with tiny caches of only 4, 2 and 6 KB for traversal, list, and transformation, respectively. Please note that shading is uncached and while all scenes use bilinear filtered textures, Scene6 and Office were rendered with unfiltered textures since otherwise the shading bandwidth would slightly limit the performance (10% at most).*

They provide insight into the hardware efficiency looking at the ratio between the number of cycles a unit was busy versus the total number of cycles it took to render the image.

Our multi-threading approach results in high usage rates. Multi-threading increases performance almost linear up to 32 threads per DynRTP. Using 64 threads further improves performance by only about 10%. However, since the resources were still available on the FPGA we used the larger number of threads for our measurements.

Even in complex scenes the external bandwidth in total is small (mostly well below 300 MB/s, ignoring the fixed 135 MB/s required for frame buffer readout due to image display at a resolution of 1024×768 with 60 Hz). The bandwidth of the DynRTC is very efficiently reduced already by tiny caches of only 12 KB total. The bandwidth requirements for shading are constant per frame as ray tracing shades every pixel exactly once and no caches are used. Only a single texture with bilinear filtering is used in the examples because the need for complex multi-texturing is greatly reduced in ray tracing as light, reflection, and environment maps are replaced by tracing the necessary rays. Bandwidth is further reduced by not having to generate these maps in the first place.

We only provide measurement data for primary rays as the performance measured in rays per second for secondary rays is identical, as it only depends on the specific arrangement of geometry in particular scenes, e.g. the placement of light sources and complexity of the scene visible through reflections. This means that switching on secondary rays can even improve the overall performance measured in rays per second.

These results confirm the results of earlier simulations [SWS02] and are obvious since the number of operations required to trace a ray and the amount of memory transfered only depends on the location of the ray in a specific scene and not on its type. The same holds for the memory bandwidth as ray coherence is mostly preserved when generating secondary rays. Secondary rays do influence the cache depending on the amount of additional data that is accessed, which again depends on the specific scene.

When using packets of rays, the performance also depends on the coherence of rays within a packet. However, coherence has been much higher than generally expected. Shadow and reflection rays are mostly coherent except for extreme cases. But even global illumination computations can be designed to use highly coherent rays [BWS03]. Our measurements on the prototype using 200 (virtual) point light sources for approximating the indirect illumination in a scene showed that the number of rays computed per second are constant compared to rendering the scene without shadow rays.

6. Conclusion and Future Work

Ray tracing is still perceived by many as an offline technique for high-quality images. Even though realtime software implementations are available for some time now, their dependence on larger clusters of PCs for good performance has been a major drawback.

With this paper we present what we believe to be the first realtime ray tracing hardware. With the prototype hardware we demonstrate that ray tracing is at least as well suited for hardware implementation as the ubiquitous rasterization approach. Even the rather simple prototype implementation of ray tracing already achieves realtime performance for a wide variety of scenes. Furthermore due to its performance which is several times higher than any current CPU, it could also be used to accellerate ray tracing in offline rendering packages.

Ray tracing hardware allows to overcome two of the main limitations of rasterization hardware. The external memory bandwidth requirements of ray tracing are only a tiny fraction compared to rasterization, where bandwidth has been a major limiting factor. Furthermore, ray tracing offers almost unlimited and very efficient scalability by adding multiple pipelines per chip, multiple chips per board, and/or multiple boards per PC.

Scalability is mainly limited by the bandwidth to the scene data. However, exactly this bandwidth can easily be reduced using packets of rays, caching, or (cached) replication of the read-only data. Ray tracing greatly benefits from its demand-driven and output-sensitive type of processing that minimizes the processing to only the relevant parts of the data.

Ray tracing hardware allows to overcome many limitations of the current rasterization approach. As demonstrated it supports huge and complex scenes, accurately computes many advanced rendering effects, and simplifies content creation significantly.

In particular we have shown that integrating a shared transformation unit into the ray tracing pipeline provides many benefits. It offers support for object-based changes to a scene by applying affine transformations to entire groups of primitives. In addition, the same unit can be re-used for greatly simplifing intersection computation as well as for implementing important parts of the shading and ray generation process.

Surprisingly, the addition of this new functionality actually *reduces* the overall hardware requirements because it factors several costly but underutilized parts of the pipeline into a single well utilized component. We were able to reduce the hardware costs by 68% compared to other implementations for dynamic ray tracers and still 25% compared to a simple static implementation with less functionality.

Of course, our prototype implementation still leaves room for many additional features and improvements. Performance seems mainly a question of getting access to more capable hardware technologies and adapting the parameters of the architecture to the capabilities of the new environment. Still, there is certainly room for extending and further improving the efficiency and performance of several parts of the current architecture.

Regarding additional features hardware ray tracing greatly benefits from the research in software implementations of realtime ray tracing. Most of the techniques developed there can be carried over to hardware with only minor changes or adaptations. This is particularly important in the context of an API for ray tracing. We believe that the OpenRT API [DWBS03] would also work well for a hardware ray tracing engine. It is currently being ported to the prototype.

Even though object-based dynamics already coveres the majority of cases, the support for dynamically changing scenes is still too limited. More flexibility is required by many applications, most notably computer games and visual simulations.

Probably the important missing feature is programmable shading, which would finally provide the full capabilities of ray tracing to applications and users. Together with the ability to easily implement global effects this seems like the killer application for ray tracing.

Finally, it remains to be seen what will be the preferred platform for realtime ray tracing in the future. Available are high-performance general purpose CPUs, large parallel programmable processing engines such as GPUs or arrays of RISC-like CPUs, or finally custom hardware. Custom hardware seems to offer the most benefits for the core ray tracing pipeline especially since it uses its floating point resources most efficiently, while other parts of ray tracing, such as shading, seem better suited for or even require general purpose-like engines. As a consequence a combination of custom hardware and more flexible engines seems like a promising approach.

Acknowledgement

We thank both Alpha-Data and Celoxica for their support of this project through providing the FPGA card. Furthermore we like to thank Ingo Wald and Carsten Benthin for their valuable contributions in our discussions.

References

[Act02] ACTIVISION: Return to Castle Wolfenstein. *http://games.activision.com/games/wolfenstein/* (2002).

[Alp03] ALPHA-DATA: ADM-XRC-II. *http://www.alphadata.uk.co* (2003).

[Are88] ARENBERG J.: Ray/Triangle Intersection with Barycentric Coordinates. *http://www.acm.org/tog/resources/RTNews/html/-rtnews5b.html* (1988).

[Bri03] BRIGHAM YOUNG UNIVERSITY, USA: BYU JHDL. *http://www.jhdl.org* (2003).

[BWS03] BENTHIN C., WALD I., SLUSALLEK P.: A Scalable Approach to Interactive Global Illumination. *Computer Graphics Forum 22*, 3 (2003), 621–630. (Proceedings of Eurographics).

[DWBS03] DIETRICH A., WALD I., BENTHIN C., SLUSALLEK P.: The OpenRT Application Programming Interface – Towards A Common API for Interactive Ray Tracing. In *Proceedings of the 2003 OpenSG Symposium*

(Darmstadt, Germany, 2003), Eurographics Association, pp. 23–31.

[Epi03] EPIC GAMES: Unreal Tournament 2003. *http://www.unrealtournament.com/ut2003/* (2003).

[Eri97] ERICKSON J.: Pluecker coordinates. *Ray Tracing News* (1997). http://www.acm.org/tog/resources/ RTNews/html/rtnv10n3.html#art11.

[GH96] GREG HUMPHREYS C. S. A.: *TigerSHARK: A Hardware Accelerated Ray-tracing Engine.* Tech. rep., Princeton University, 1996.

[GP89] GREEN S. A., PADDON D. J.: Exploiting coherence for multiprocessor ray tracing. *IEEE Computer Graphics and Applications 9*, 6 (1989), 12–26.

[GP90] GREEN S. A., PADDON D. J.: A highly flexible multiprocessor solution for ray tracing. *The Visual Computer 6*, 2 (1990), 62–73.

[Gre91] GREEN S. A.: Parallel processing for computer graphics. *MIT Press* (1991), 62–73.

[Hal01] HALL D.: The AR350: Today's ray trace rendering processor. In *Proceedings of the Eurographics/SIGGRAPH workshop on Graphics hardware - Hot 3D Session 1* (2001).

[Id-99] ID-SOFTWARE: Quake3-Arena. *http://www.quake3arena.com/* (1999).

[KH95] KEATES M. J., HUBBOLD R. J.: Interactive ray tracing on a virtual shared-memory parallel computer. *Computer Graphics Forum 14*, 4 (1995), 189–202.

[KiSSO02] KOBAYASHI H., ICHI SUZUKI K., SANO K., OBA N.: Interactive Ray-Tracing on the 3DCGiRAM Architecture. In *Proceedings of ACM/IEEE MICRO-35* (2002).

[LAM00] LEXT J., ASSARSSON U., MÖLLER T.: *BART: A Benchmark for Animated Ray Tracing.* Tech. rep., Department of Computer Engineering, Chalmers University of Technology, Göteborg, Sweden, May 2000. Available at http://www.ce.chalmers.se/BART/.

[LAM01] LEXT J., AKENINE-MÖLLER T.: Towards Rapid Reconstruction for Animated Ray Tracing. In *Eurographics 2001 – Short Presentations* (2001), pp. 311–318.

[LS91] LIN T. T., SLATER M.: Stochastic Ray Tracing Using SIMD Processor Arrays. *The Visual Computer* (1991), 187–199.

[MKS98] MEISSNER M., KANUS U., STRASSER W.: VIZARD II, A PCI-Card for Real-Time Volume Rendering. In *Eurographics/Siggraph Workshop on Graphics Hardware* (1998).

[MPJ*00] MAI K., PAASKE T., JAYASENA N., HO R., DALLY W., HOROWITZ M.: Smart Memories: A Modular Recongurable Architecture. *IEEE International Symposium on Computer Architecture* (2000).

[MT97] MOELLER T., TRUMBORE B.: Fast, minimum storage ray triangle intersection. *Journal of Graphics Tools 2*, 1 (1997), 21–28.

[Muu95] MUUSS M. J.: Towards real-time ray-tracing of combinatorial solid geometric models. In *Proceedings of BRL-CAD Symposium '95* (June 1995).

[Nvi04] NVIDIA: GeForce FX. *http://www.nvidia.com* (2004).

[PBMH02] PURCELL T. J., BUCK I., MARK W. R., HANRAHAN P.: Ray Tracing on Programmable Graphics Hardware. In *Proceedings of SIGGRAPH 2002* (2002).

[Pfi01] PFISTER H.-P.:. SIGGRAPH course on Interactive Ray Tracing, 2001.

[PHK*99] PFISTER H., HARDENBERGH J., KNITTEL J., LAUER H., SEILER L.: The VolumePro real-time raycasting system. *Computer Graphics 33* (1999).

[PSL*99] PARKER S., SHIRLEY P., LIVNAT Y., HANSEN C., SLOAN P. P.: Interactive ray tracing. In *Interactive 3D Graphics (I3D)* (April 1999), pp. 119–126.

[Pur01] PURCELL T.: The SHARP Ray Tracing Architecture. SIGGRAPH course on Interactive Ray Tracing, 2001.

[Pur04] PURCELL T. J.: *Ray Tracing on a Stream Processor.* PhD thesis, Stanford University, 2004.

[RSH00] REINHARD E., SMITS B., HANSEN C.: Dynamic Acceleration Structures for Interactive Ray Tracing. In *Proceedings of the Eurographics Workshop on Rendering* (Brno, Czech Republic, June 2000), pp. 299–306.

[SLS03] SCHMITTLER J., LEIDINGER A., SLUSALLEK P.: A Virtual Memory Architecture for Real-Time Ray Tracing Hardware. *Computer and Graphics, Volume 27, Graphics Hardware* (2003), 693–699.

[SWS02] SCHMITTLER J., WALD I., SLUSALLEK P.: SaarCOR – A Hardware Architecture for Ray Tracing. In *Proceedings of the ACM SIGGRAPH/Eurographics Conference on Graphics Hardware* (2002), pp. 27–36.

[Wal04] WALD I.: *Realtime Ray Tracing and Interactive Global Illumination.* PhD thesis, Computer Graphics Group, Saarland University, 2004. Available at http://www.mpi-sb.mpg.de/~wald/PhD/.

[WBS03] WALD I., BENTHIN C., SLUSALLEK P.: Distributed Interactive Ray Tracing of Dynamic Scenes. In *Proceedings of the IEEE Symposium on Parallel and Large-Data Visualization and Graphics (PVG)* (2003).

[WBWS01] WALD I., BENTHIN C., WAGNER M., SLUSALLEK P.: Interactive Rendering with Coherent Ray Tracing. *Computer Graphics Forum (Proceedings of EUROGRAPHICS 2001 20*, 3 (2001).

[WPS*03] WALD I., PURCELL T. J., SCHMITTLER J., BENTHIN C., SLUSALLEK P.: Realtime Ray Tracing and its use for Interactive Global Illumination. In *Eurographics State of the Art Reports* (2003).

[Xil03] XILINX: Virtex-II. *http://www.xilinx.com* (2003).

Graphics Hardware (2004)
T. Akenine-Möller, M. McCool (Editors)

A Programmable Vertex Shader with Fixed-Point SIMD Datapath for Low Power Wireless Applications

Ju-Ho Sohn[†] Ramchan Woo[†] Hoi-Jun Yoo[‡]

Semiconductor System Laboratory, Department of Electrical Engineering and Computer Science,
Korea Advanced Institute of Science and Technology (KAIST), Daejeon, Korea

Abstract

The real time 3D graphics becomes one of the attractive applications for 3G wireless terminals although their battery lifetime and memory bandwidth limit the system resources for graphics processing. Instead of using the dedicated hardware engine with complex functions, we propose an efficient hardware architecture of low power vertex shader with programmability. Our architecture includes the following three features: I) *a fixed-point SIMD datapath to exploit parallelism in vertex processing while keeping the power consumption low,* II) *a multithreaded coprocessor interface to decrease unwanted stalls between the main processor and the vertex shader, reducing power consumption by instruction-level power management,* III) *a programmable vertex engine to increases the datapath throughput by concurrent operations with main processor. Simulation results show that full 3D geometry pipeline can be performed at 7.2M vertices/sec with 115mW power consumption for polygons using the OpenGL lighting model. The improvement is about 10 times greater than that of the latest graphics core with floating-point datapath for wireless applications in terms of processing speed normalized by power consumption, Kvertices/sec per milliwatt.*

Category and Subject Descriptors: I.3.1 [Computer Graphics]: Hardware Architecture—Graphics Processor C.1.2 [Processor Architecture]: Multiple Data Stream Architectures—Single–instruction–stream, multiple–data–stream processor (SIMD)

1. Introduction

As the mobile electronics market increases rapidly, 3G wireless terminals such as PDAs or mobile phones get more popular. Since these applications have the display devices, the rendering capabilities from simple two-dimensional graphics even to the real time three-dimensional graphics will be necessary to provide more realistic functionalities.

For the wireless applications, the low power consumption is the most important issue because they are driven by batteries. The Advanced RISC Machines (ARM) processor family that has the reduced instruction set computer (RISC) architecture is most widely used as a main platform for the wireless applications because of its high MIPS/Watt [Cla02]. Since there are very limited system resources in terms of computation power and memory bandwidth, the graphics architecture should be designed to consume as little energy as possible on the ARM platform. Moreover, since the users hold the small screens of wireless terminals close to their eyes, the average eye-to-pixel angle is wider than that of a PC system. Therefore we must provide more realistic graphics images with low power consumption in wireless terminals.

Recently, several researches have tried to increase the mobile graphics capabilities in wireless applications. Since the rasterization and texture mapping require more processing complexities than the rest of operations in the 3D graphics pipeline [ODK*00], most of graphics architectures have mainly focused on the rendering pipeline and achieved the efficient graphics performances [AMS03][WCS*03]. However, since relatively little attention has been given to 3D geometry operations, now they become the performance barriers in 3D graphics pipeline. The geometry units of graphics systems in PC and workstations [LKM01][MBD*97] show high performance while consuming much power. For wireless applications, the general-purpose RISC processors with simple integer datapath [WCS*03] or conventional floating-point datapath [KKF*03] were used to process vertex shading operations. However, the simple integer

[†] {sohnjuho, ural}@eeinfo.kaist.ac.kr
[‡] hjyoo@ee.kaist.ac.kr

datapath could not provide the required performance of vertex shading. In conventional floating-point datapath, the performance is also limited due to low operating frequency within allowed budget of limited power consumption.

In this paper, we propose a programmable vertex shader for the low power wireless applications. It can fill the gap between the flexible high performance 3D geometry systems and the low power wireless platforms with limited system resources. The proposed hardware architecture for low power vertex shader has three major features:

1) a fixed-point SIMD datapath that uses an energy-efficient fixed-point arithmetic to exploit data level parallelism in vertex processing while keeping the power consumption low,

2) a multithreaded ARM coprocessor interface - an instruction extension mechanism of ARM platform, which decreases the unwanted stalls between the main processor and the vertex shader, and reduces power consumption by instruction-level power management, and

3) a programmable vertex engine that increases the datapath throughput for streaming vertex data input by concurrent operations with the main processor while saving the external memory bandwidth.

This paper is organized as follows. First, the hardware architecture of the proposed vertex shader is described, and followed by a discussion about the program model. Then the implementation details are presented. After that, simulation results and evaluations are provided. Finally conclusions are made.

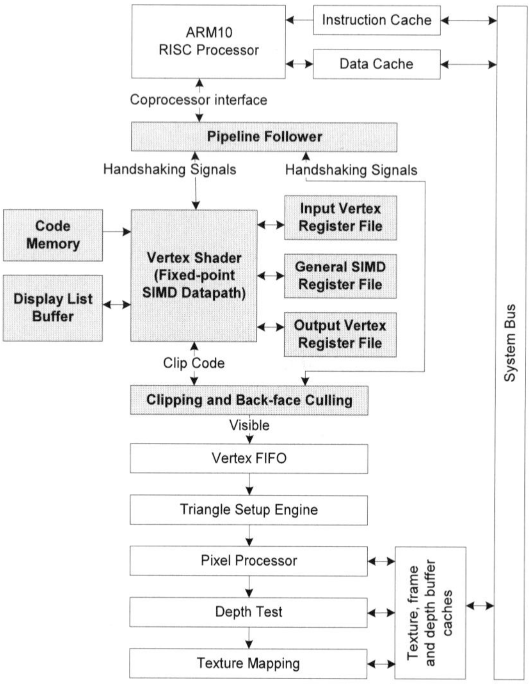

Figure 1: System configuration and block diagram of proposed vertex shader

2. Architecture

Figure 1 shows the system configuration of our graphics architecture and the gray blocks indicate the proposed vertex shader. All hardware blocks including vertex shader are connected to the ARM10 RISC processor through the coprocessor interface, and use system bus interfaces for external memory accesses. In this section, the hardware architecture of the vertex shader is described focusing the three previously mentioned features.

2.1. Fixed-point SIMD Datapath

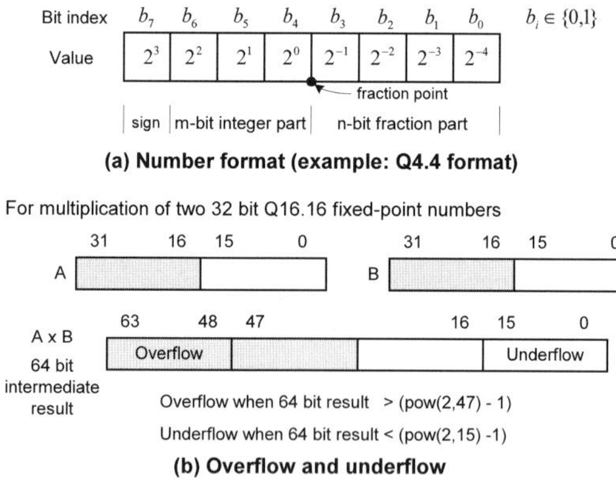

Figure 2: Fixed-point representation

Most 3D graphics systems require real number representation to support various 3D rendering algorithms. We use fixed-point number format for real number representation as shown in Figure 2.(a) [Kol02] instead of using floating-point number. There are two reasons for this. First, the hardware architecture of fixed-point arithmetic is much simpler than that of floating-point arithmetic because fixed-point arithmetic uses only integer datapath. Therefore the fixed-point unit can occupy less silicon area, and it can consume less power. Second, because fixed-point unit can operate at higher clock speed, it can process the same task faster than floating-point unit. Therefore, we can use less energy when applying fixed-point arithmetic in the graphics operations. For matrix multiplication, 32-bit fixed-point hardware consumes 17% less power than single precision floating-point multiplier. It can also operate at 30% higher clock frequency.

To evaluate the accuracy of the fixed-point arithmetic in the 3D geometry operations, the following equations can be used to decide the number of bits for fractional part of Qm.n fixed-point number [HV01], where the 'm' is the number of bits representing integer part and 'n' is the number of bits representing fractional part.

for transformation,

$$n_f = n_a + 3 + \left\lceil \log_2\left(1 + \frac{\text{distance of far plane from eye}}{\text{distance of scene vertex to eye}}\right) \right\rceil$$

for lighting,

$$n_f = n_a + 8 \quad \text{or} \quad n_a + 9$$

where 'n_f' is the number of fractional bits to get n_a bits of accuracy after transformation and lighting calculations.

The screen resolution and color depth of wireless terminals are relatively small, for example the common displays of today's wireless application such as QVGA has 320 x 240 resolution with 16-bit color depth. In this case, 14 or 15 bits are enough for the fractional part to get pixel level accuracy

Our architecture utilizes 128-bit wide 4-way SIMD instructions, enabling to concurrently process up to four 32 bit fixed-point data elements in a single cycle. To improve the usefulness in fixed-point arithmetic, status registers are applied to indicate the overflows and underflows occurred in the multiplications of two fixed-point numbers as shown in Figure 2.(b). These status registers can be used to check errors in fixed-point arithmetic without extra cycle penalties.

In order to enhance the dynamic range further in real number representation, we add controlled ADD/SUB (CAS) instruction and controlled logical shift (CLS) instruction (Figure 3) for software emulation of floating-point arithmetic to fixed-point SIMD instructions. After updating negative flag in arithmetic status register by previous instructions such as SUB, we can make CAS instruction be a single ADD instruction or SUB instruction. We can also make CLS instruction be a single left shift instruction or right shift instruction. These instructions can reduce the unnecessary comparison operations in exponent alignment and normalization of floating-point arithmetic. With floating-point emulation, our architecture shows 160Mflops performance at 400MHz clock speed.

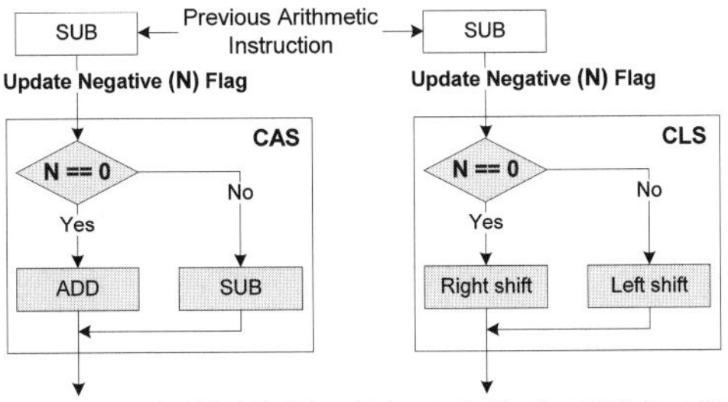

(a) Controlled Add/Sub (CAS) (b) Controlled Left shift/Right shift (CLS)

Figure 3: Special instructions for floating-point emulation

2.2. Multithreaded ARM Coprocessor

The ARM architecture supports an extension of the instruction set architecture by adding coprocessors [Fur02]. The proposed architecture is integrated with ARM10 RISC processor. The coprocessor interface is used to interconnect the proposed architecture with main processor because of the following reasons

- Since the coprocessor operates in lock step with core pipeline of the main processor, it can avoid a complex synchronization and provide a single context of control.
- The coprocessors have the direct signal paths from the main processor. They don't need the bus arbitrations for hardware accesses contrary to conventional hardware accelerators, which use the shared system bus. Therefore, coprocessor interface can reduce the unwanted stalls between main processor and hardware accelerators.
- Since the data cache of main processor can be shared to store graphics primitives as well, additional memories are not necessary to store vertex data.

To achieve high graphics performance at low power consumption, we add two additional features to the conventional coprocessor. First, a multithread architecture is proposed to enable the stream processing of vertex data. For this, three independently accessible register files are used for storing input vertex data, output vertex data and temporary results. The vertex shader processes polygon data stored in input vertex register file, and generates the shaded vertex output to the output vertex register file. The general purpose SIMD register file is used to hold the temporary results. Second, the coprocessor can be activated only when it is required on an instruction-by-instruction basis. If the current instruction is not a valid

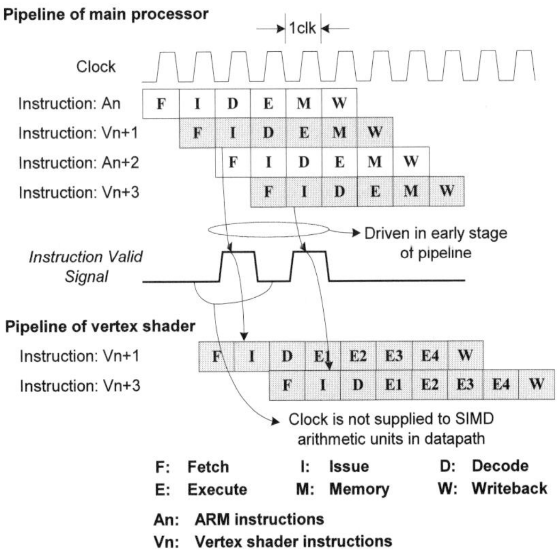

Figure 4: Instruction pipeline and clock gating

coprocessor instruction, clock signal is not supplied to the fixed-point SIMD arithmetic units in datapath. As shown in Figure 4, the instruction valid signal driven in the early stage of the main processor pipeline disables the unnecessary hardware blocks of the coprocessor datapath to reduce power consumption.

2.3. Programmable Vertex Engine

In general, the ARM coprocessor cannot be operated without instruction issuing from the main processor. Therefore, the datapath of the main processor is stalled while the coprocessor executes instructions. Because it leads to the performance degradation of main processor when processing the streaming vertex data, we improve the throughput of main processor and vertex shader by allowing coprocessor to execute the vertex program independently.

Figure 5 shows the internal architecture of the vertex engine. After downloading the vertex program into the internal code memory via the coprocessor interface, the vertex shader starts executing each instruction running free from main processor. As previously mentioned, the vertex shader is a multithreaded coprocessor with fixed-point SIMD datapath. The input vertex register file that is used to hold the vertex attributes such as position and normal vector is fed into the fixed-point SIMD datapath. The vector arithmetic unit is responsible for all arithmetic operations such as addition and multiplication, and the special function unit is responsible for reciprocal (RCP) and reciprocal square root (RSQ). Most of the operations are performed as the 32 bit fixed-point numbers, and achieve a single cycle pipelined throughput. The constants such as transformation matrix, lighting parameters and lookup table entries are stored in the display list buffer and temporary results are stored in the general purpose SIMD register file. The shaded vertex output is transformed into one of the ouput vertex register files. There are three output vertex reigter files for caching of vertex data in the primitive assembly and only one of them is accessible in the vertex program. To save the silicon area, only one input vector operand can be swizzled aribitrarily in the SIMD datapath and all the output writes can be controlled by component-wise write mask bits

In the vertex shader, we have the display list buffer to store graphics primitives, reducing the traffic on external memory I/O. Also, the display list buffer can be shared to hold vertex constants at the same time for design simplicity of hardware. To enhance the efficiency of addressing, we have two integer address registers for indexed display list buffer reads. In addtion, the display list buffer has the following two features.

▪ Auto increment and decrement addressing modes: the address register can be updated automatically after indexed display list buffer reads, which is useful to manage the vertex streams.

▪ 8 bit / 16 bit unpack with shuffling of vector components: For geometry compression, the 8 bit or 16 bit read data from the display list buffer can be automatically sign-extended to the full 32 bit fixed-point numbers, which can be used as the delta difference between one vertex and the next vertex [Dee95].

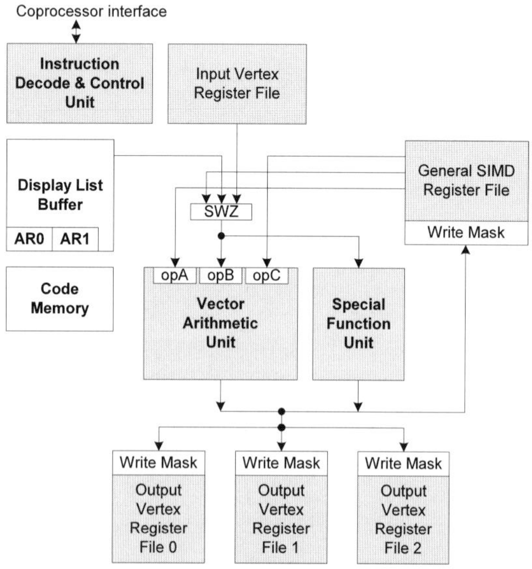

Figure 5: Internal architecture of vertex engine

3. Program Model

In this section, we will describe the program model of the proposed vertex shader including the processor operating states and the instruction set architecture.

3.1. Dual Personality

The main purpose of our design is to use the simple program model through ARM coprocessor interface, while improving the datapath performance with low power consumption for vertex data streams. From the programmer's point of view, the vertex shader can be in one of two states as shown in Figure 6.

1) Tightly coupled coprocessor (TCC): All instruc-

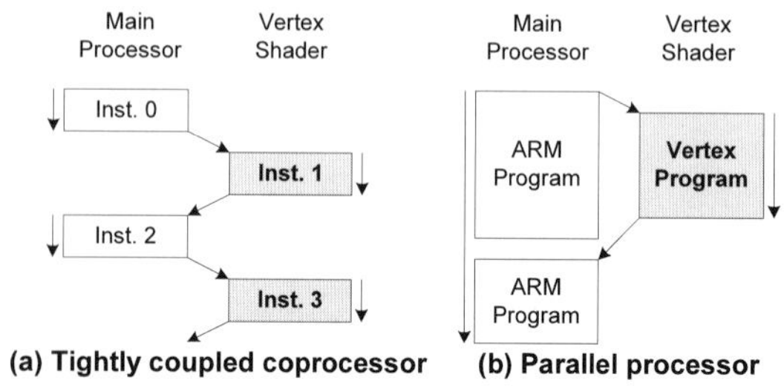

Figure 6: Processor operating states

tions of the vertex shader are issued in the instruction stream of the main processor as extended instructions and, they are executed in lock step with main pipeline.

2) Parallel processor (PP): In this state, vertex shader can process the vertex program stored in the internal code memory without instruction issues from main processor. The vertex shader can achieve the maximum throughput of datapath because it can operate concurrently with main processor [Gan98]. The main processor reads the vertex attributes in the main memory and feeds them into the vertex shader while vertex program is executed. To maintain the communication protocol of the ARM coprocessor interface, the vertex shader drives the coprocessor busy signals to the main processor, leading next coprocessor instruction to busy-wait loop for system synchronization.

3.2. Instruction Set

We use two kinds of instruction set architectures for each processor operating state. The instruction set in TCC state contains all data processing and transfer instructions for vertex shader. In PP state the instruction set consists of 20 operations, which are the modified subset of the previous programmable vertex engine [LKM01]. Table 1 shows the instruction set of vertex program that can be executed in PP state.

Opcode	Full name	Description	Latency	Throughput
MUL	Multiply	Vector → Vector	4	1
MAD	Multiply and add	Vector → Vector	4	1
DP3	3 term dot product	Vector → Replicated scalar	5	2
DP4	4 term dot product	Vector → Replicated scalar	5	2
TRFM	Transform	Vector → Vector	7	4
ADD	Addition	Vector → Vector	1	1
SUB	Subtraction	Vector → Vector	1	1
MOV	Move	Vector → Vector	1	1
RCP	Reciprocal	Scalar → Replicated scalar	6	3
RSQ	Reciprocal square root	Scalar → Replicated scalar	8	5
MIN	Minimum	Vector → Vector	1	1
MAX	Maximum	Vector → Vector	1	1
SLT	Set on less than	Vector → Vector	1	1
SGE	Set on grater or equal	Vector → Vector	1	1
SEQ	Set on equal	Vector → Vector	1	1
LSL	Logical shift left	Vector → Integer Vector	1	1
ASR	Arithmetic shift right	Vector → Integer Vector	1	1
ZERO	Set zero	Vector	1	1
ARL	Address register load	Vector → Scalar integer	2	2
END	Vertex program end	Miscellaneous	1	1

Table 1: Instruction set of vertex program

3.2.1 Control flow instructions

In TCC state, the control flow instructions such as branch and return are managed by the main processor and the vertex shader executes only the extended SIMD arithmetic instructions. However all vertex shader instructions are conditionally executed to maximize execution throughput. They affect on the memory and registers only if the arithmetic flags (negative, zero, carry out and overflow) satisfy a condition specified in the instruction. When implementing the fixed function pipeline of graphics library such as OpenGL which is controlled by global states, the state checking, vertex shading path selection, homogeneous clip space operations and back face culling are handled in TCC state. The remaining code segments for actual vertex shading operations can be executed without state checking. These operations are carried out in the vertex program of PP state. It supports the vertex transform paths without branching for simplicity and efficiency of hardware architecture. Even if the control flow instructions are not supported in PP state, simple if/then/else statement is still possible through SLT, SGE and SEQ instructions.

3.2.2 Fixed-point SIMD arithmetic instructions

To save the system resources, we make the datapath simple and efficient without complex hardware blocks. All arithmetic operations including RCP and RSQ are executed on the fixed-point numbers that can have any precisions, using only low power interger arithmetic units. They are also fully pipelined and bypassed to forward the data to the different stage of pipeline of correct instructions.

Since the multiplication equivalent instructions consume most of time in vertex shading operations, we make the throughput of MUL and MAD be a single cycle, and support dot products for coding convenience. Moreover we add TRFM instruction to enhance the performance of the vertex transformation in

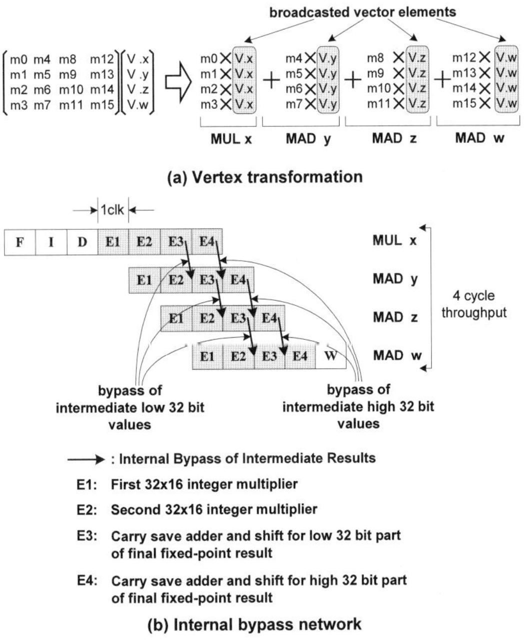

Figure 7: TRFM instruction

homogeneous coordinates. Although the latency of MUL and MAD are four cycles, TRFM instruction can calculate the vertex transform-ation at every four cycles with the help of broadcasting of vector elements and internal bypass network of intermediate values as shown in Figure 7.

For design simplicity, we remove the complex functions such as the logarithmic, exponential and specular power functions, and rather the table look-up is used for these functions. The integer shift instructions of fixed-point numbers are added in order to extract bit fields for index calculations in the lookup table. After shift operations of vertex specific index, the ARL instruction can allow an offset into the lookup table.

The following vertex program implements the vertex transformation and full Phong shading. It uses OpenGL lighting equations assuming infinite light and viewpoint positions. To calculate the specular power function, we use the lookup table of 64 entries which store the specular coefficients for given shininess value. After calculating dot product of normal vector and the Blinn halfway vector, we use the integer shift instructions for offset values. After the rearrangements of the instructions for compiler optimization, the proposed hardware running at 400MHz can process these vertices at a rate of 7.2M vertex/sec including view frustum clip check, perspective division and viewport transform.

```
# Vertex Transformation and OpenGL Lighting
#
# c[0-3]    = modelview matrix (column-wise)
# c[4-7]    = modelview inverse transpose (column-wise)
# c[8-11]   = modelview projection matrix (column-wise)
# c[16]     = light position
# c[17]     = blinn halfway vector
# c[18]     = precomputed specular light * specular mat.
# c[19]     = precomputed diffuse light  * diffuse mat.
# c[20]     = precomputed ambient light  * ambient mat.
# c[32-47]  = 64 entries lookup table for specular power (column-wise)
# c[48]     = 16th, 32th, 48th and 64th entries of lookup table
# c[49].x   = fraction bit length of fixed-point format
# c[49].y   = fraction bit length - 2
# c[49].z   = fraction bit length - 6
# c[49].w   = fraction bit length + 4
# c[50]     = (0, 1, 2, 3) in integer format

# Vertex transformation to eye space
TRFM VGR0.xyz, VIR[OPOS], c[0];

# Normal vector transform to eye space
TRFM VGR1.xyz, VIR[NRML], c[4];

# Vertex transformation to clip space
TRFM VOR[HPOS], VIR[OPOS], c[8];

# Compute normalized light direction
SUB  VGR0.xyz, VGR0, c[16];
DP3  VGR0.w, VGR0, VGR0;
RSQ  VGR0.w, VGR0.w;
MUL  VGR0.xyz, VGR0, VGR0.w;

# Compute N.L and N.H
DP3  VGR2, VGR1, VGR0;
DP3  VGR3, VGR1, c[17];

# Index calculation of lookup table for specular power function
ASR  VGR4, VGR3, c[49].y;
ASR  VGR5, VGR3, c[49].z;
LSL  VGR6, VGR5, c[49].z;
SUB  VGR6, VGR3, VGR6;
LSL  VGR5, VGR5, c[49].x;
LSL  VGR7, VGR4, c[49].w;
SUB  VGR5, VGR5, VGR7;

# table look-up
ARL  A0.x, VGR5.x;
SEQ  VGR7.x, VGR4, c[50].x;
SEQ  VGR7.y, VGR4, c[50].y;
SEQ  VGR7.z, VGR4, c[50].z;
SEQ  VGR7.w, VGR4, c[50].w;
DP4  VGR4, VGR7, c[A0.x+32];
DP4  VGR5, VGR7, c[A0.x+33];

# Compute specular power using interpolation
SUB  VGR5, VGR5, VGR4
MAD  VGR3, VGR5, VGR6, VGR4;

# Compute light color values
MUL  VGR5.xyz, VGR3, c[18];
MUL  VGR4.xyz, VGR2, c[19];
ADD  VGR5.xyz, VGR4, VGR5
ADD  VOR[COL0].xyz, VGR5, c[20];

# texture coordinate
MOV  VOR[TEX0], VIR[TEX0];
```

4. Implementation

We have developed a fully synthesizable Verilog-HDL model of the proposed graphics architecture including vertex shader and a hardware rasterizer. The vertex shader consists of 230K logic gates and 280Kbit SRAM using 0.13um process. The capacity of display list buffer is 32KByte. The low power rendering engine integrated in our previous work [WCS*03] was applied as a rasterizer. For system evaluations, we designed a software graphics library, a subset of OpenGL, which utilizes the fixed-point arithmetic operations to optimally use the features of implemented hardware architecture. To manage the vertex shader such as downloads of vertex programs and vertex attributes, simple API calls are provided as OpenGL extension. After implementing software test bench using synthesized gate level circuits, we measured the vertex processing speed and power consumptions.

5. Results and Evaluation

In this section, we show hardware simulation results focusing the processing speed and power consumption.

In order to measure the accuracy of fixed-point arithmetic in vertex shading operations, we compare the rendered images of 3D objects using software only floating-point graphics library with our hardware using fixed-point graphics library. In Figure 8, lit, smooth-shaded spheres with different material properties and an animated 3D character are rendered to show reliability of the proposed vertex shader. From the results, we can show that under vertex-level accuracy, the maximum transformed distance between floating-point and fixed-point systems is less than 0.000025 for Q12.20 fixed-point format and 0.0002 for Q16.16 fixed-point format.

Figure 9 shows vertex processing speed and power consumption in case of unlit, diffuse lit and diffuse-specular lit polygons. For diffuse-specular lit polygons, full 3D geometry pipeline can be performed at a rate of 7.2M vertices/sec with 115mW power consumption

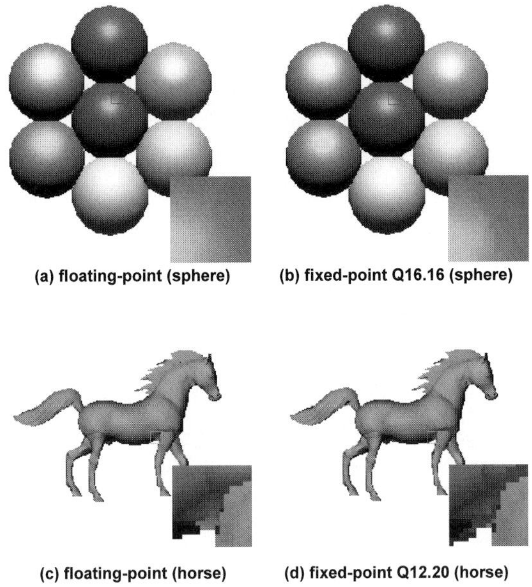

Figure 8: Fixed-point vertex shading of 3D objects (spheres: 5068 vertices, horse: 6798 vertices)

including host processor, vertex shader and display list buffer. In other cases, the performance was improved with less power consumption because unnecessary hardware blocks such as special function unit and lookup table can be disabled.

Figure 10 shows the relationship between rendered vertex counts and processing speed with various sizes of the vertex data format. The vertex processing speed was measured with and without using the display list buffer for vertex caching. To reduce the external memory bandwidth, vertex data in the display list buffer were stored as a 8 bit or 16 bit data type. In typical wireless applications, a 32-bit SDRAM memory running at 100 MHz is used as the main memory, and gives a usable bandwdith of 200MB/s with approximately 50% of efficiency. We restriced the peak bandwidth requirement

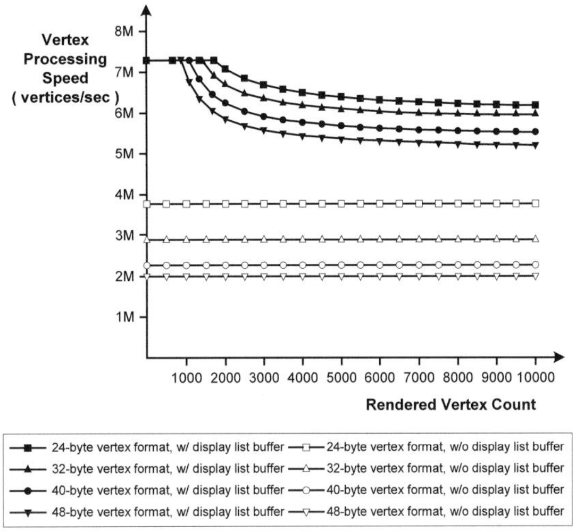

Figure 10: Relationship between rendered vertex counts and processing speed

of the vertex shader to 100MB/s in order to allocate the remaining bandwidth for other devices. As shown in this figure, the vertex shader shows twice higher performance when using the display list buffer than only data cache of the main processor. Moreover, the performance degradation is merely 10% as vertex counts are increased. Because we can feed vertex data stream continuously during execution of vertex program with the help of concurrent operations of the main processor, we can achieve high sustaining performance.

With high vertex processing speed, the hardware vertex shader in the PC and workstation platforms provide many advanced shading functions. However they consume a great deal of power, more than several tens of watts. Therefore we should use the following performance indices to measure the performance of vertex shader in wireless applications taking into account the power consumption [WYK*02].

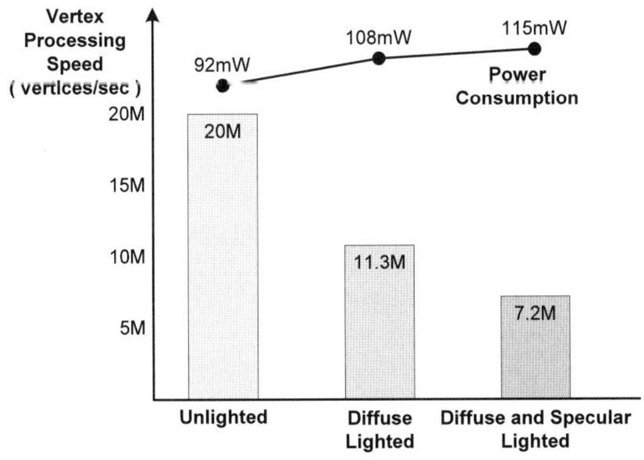

Figure 9: Processing speed and power consumption

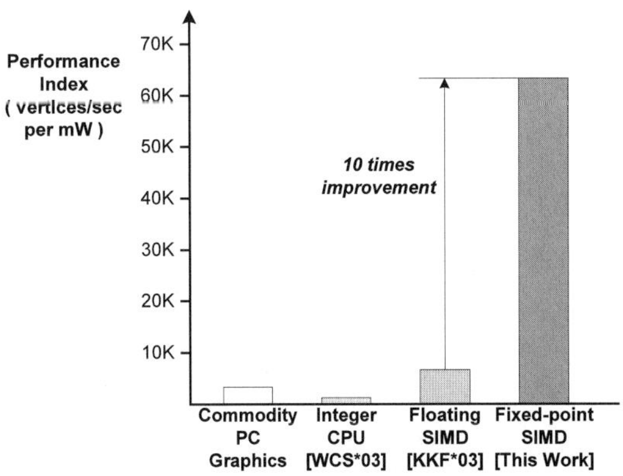

Figure 11: Performance comparison

$$\text{Vertex Shading Performance in Wireless Applications} = \frac{\text{Vertex Processing Speed (vertices/sec)}}{\text{Power Consumption in Vertex Shader(mW)}}$$

Figure 11 shows the performance comparison of proposed vertex shader with different graphics architecture in terms of Kvertices/sec per milliwatt. The improvement is about 10 times greater than that of the latest graphics core with floating-point datapath for wireless applications [KKF*03]. In comparison with simple integer datapath of the conventional low power embedded CPU [WCS*03], it shows far higher performance improvement.

6. Conclusions

We have presentated the design and implementation of a programmable vertex shader with fixed-point SIMD datapth for low power wireless applications. Most graphics architectures for wireless applications have mainly focused on rasterization and texture mapping due to high processing requirements. In order to balance 3D graphics pipeline within the limited system resources, we used simple and efficient programmable architecture for vertex shading instead of using dedicated hardware engine with complex functions. Since main purpose of the proposed design is to provide high performance with low power consumption, we used a energy-efficient fixed-point SIMD datapath to exploit parallelism in vertex shading operations. It shows comparable image quality to floating-point system, while operating at higher clock speed with low power consumption. For simple program model, the proposed vertex shader is connected to main processor through dedicated ARM coprocessor interface. Moreover, using multithredaded architecture, it can handle vertex data stream more efficiently. In addition to conventional ARM coprocessor, our architecture can be operated as a programmable vertex engine and execute vertex program concurrently with the main processor. All instructions are optimized to fixed-point arithmetic.

The architecture has been implemented in fully syntheiszable hardware model. Simulation results show that full 3D geometry pipeline can be performed at a rate of 7.2M vertices/sec with 115mW power consumption for polygons using the OpenGL lighting model. The improvement is about 10 times greater than that of the latest graphics core with floating-point datapath for wireless applications in terms of processing speed normalized by power consumption, or vertices/sec per milliwatt.

In the future, we would like to enhance the fixed-point artihmetic in vertex shading operations, and combine our architecture with low power programmable pixel shader to provide more higher quality of rendering images for low power wireless applications.

References

[AMS03] Akenine-Möller T., Ström J.: Graphics for the masses: A hardware rasterization architecture for mobile phones. *In Proc. SIGGRAPH 2003, pp. 801-808, July, 2003*

[Cla02] Clark D.: Mobile processors begin to grow up. *IEEE Computer, vol. 35, Issue 3, pp. 22-24, March, 2002*

[Dee95] Deering M.: Geometry compression. *In Proc. SIGGRAPH 95, pp. 13-20, July, 1995*

[Fur02] Furber S.: ARM: System-on-chip architecture. *2nd edition Addison-Wesley Press, 2000*

[Gan98] Gandhi P.: SA1500:A 300MHz RISC CPU with attahced media processor. *In Proc. Hot Chips 10, August, 1998*

[HV01] Hao X., Varshney A.: Variable-precision rendering. *In Proc. the 2001 Symp. on Interative 3D Graphics, pp. 149-158, 2001*

[KKF*03] Kameyama M., Kato Y., Fujimoto H., Negishi H., Kodama Y., Inoue Y., Kawai H.: 3D graphics LSI core for mobile phone –Z3D. *In Proc. SIGGRAPH/Eurogrphics Workshop on Graphics Hardware 2003, pp.60-67, August, 2003*

[Kol02] Kolli G. K.: 3D graphics optimization for ARM architecture. *presented at the Game Developer Conf., San Jose, CA. 2002*

[LKM01] Lindholm E., Kilgard M., Moreton H.: A user-programmable vertex engine. *In Proc. SIGGRAPH 2001, pp. 149-158, August, 2001*

[MBD*97] Montrym, J. S., Baum D. R., Dignam D. L., Migdal C. J.: Infinite Reality: A Real-Time Graphics System. *In Proc. SIGGRAPH 1997, pp. 293-302, August, 1997*

[ODK*00] Owens J. D., Dally W. J., Kapasi U. J., Rixner S., Mattson P., Mowery B.: Polygon rendering on a stream architecture. *In Proc. SIGGRAPH/Eurographics Workshop on Graphics Hardware 2000, pp. 23-32, August, 2000*

[WCS*03] Woo R., Choi S., Sohn J., Song S., Bae Y., Yoo H.: A low power and high performance 2D/3D graphics accelerator for mobile multimedia applications. *In Proc. Hot Chips 15, August, 2003*

[WYK*02] Woo R., Yoon C., Kook J., Lee S., Yoo H.: A 120mW 3D rendering engine with a 6-Mb embedded DRAM and 3.2 GB/s run-time reconfigurable bus for PDA chip. *IEEE J. of Solid-State Circuits, vol. 3, No. 10, pp.1352-1355, October, 2002*

Graphics Hardware (2004)
T. Akenine-Möller, M. McCool (Editors)

UberFlow: A GPU-Based Particle Engine

Peter Kipfer, Mark Segal, Rüdiger Westermann

Computer Graphics & Visualization, Technische Universität München† , ATI Research‡

Abstract

We present a system for real-time animation and rendering of large particle sets using GPU computation and memory objects in OpenGL. Memory objects can be used both as containers for geometry data stored on the graphics card and as render targets, providing an effective means for the manipulation and rendering of particle data on the GPU.
To fully take advantage of this mechanism, efficient GPU realizations of algorithms used to perform particle manipulation are essential. Our system implements a versatile particle engine, including inter-particle collisions and visibility sorting. By combining memory objects with floating-point fragment programs, we have implemented a particle engine that entirely avoids the transfer of particle data at run-time. Our system can be seen as a forerunner of a new class of graphics algorithms, exploiting memory objects or similar concepts on upcoming graphics hardware to avoid bus bandwidth becoming the major performance bottleneck.

Categories and Subject Descriptors (according to ACM CCS): I.3.7 [Computer Graphics]: Three-Dimensional Graphics and Realism

1. Introduction

From a conceptual point of view, a particle engine consists of a set of connected components or modules responsible for the creation, manipulation and rendering of particle primitives [Ree83, LT93, BW97, McA00, Sim90]. In computer graphics, particle engines are most commonly used to generate volumetric effects like fire, explosions, smoke or fluids, based on simple physical models. More elaborate particle engines are employed for physically based dynamics simulation (e.g. Smoothed Particle Hydrodynamics [Mon88], Boltzmann models [Boo90, Doo90]). Such systems usually integrate particle-particle interactions as well as more complex physical models to describe particle dynamics.

In current particle engines, the dynamics module is run on the CPU. The rendering module sends particle positions and additional rendering attributes to the GPU for display. This conventional assignment of functional units to processing units reveals the capabilities of early generations of graphics processors. Such processors were solely optimized for the *rendering* of lit, shaded and textured triangles. Nowadays, this design is abandoned in favor of programmable function pipelines that can be accessed via high level shading languages [MGAK03, Mic02]. On current GPUs, fully programmable parallel geometry and fragment units are available providing powerful instruction sets to perform arithmetic and logical operations. In addition to computational functionality, fragment units also provide an efficient memory interface to server-side data, i.e. texture maps and frame buffer objects.

Both in conventional particle engines and in engines that exploit parallel computations and memory bandwidth on the GPU, updated particle positions have to be transferred to client memory before they can be used as input for the geometry engine. Then, bandwidth is becoming a major performance bottleneck, and with the ability to do more operations per time interval on the CPU or the GPU, the bandwidth required will grow substantially. Consequently, for high resolution particle sets the transfer of these sets for rendering purposes has to be avoided.

OpenGL memory objects and similar concepts, i.e. texture access in vertex shaders using PixelShader 3.0, provide this functionality. A memory object is a block of data in graph-

† kipfer@in.tum.de, westermann@in.tum.de
‡ segal@ati.com

ics memory that can be used simultaneously as vertex buffer and render target. This mechanism allows one to direct the output of a fragment program into such objects, and to feed the data into the geometry engine as an array of vertex positions or additional per-vertex attributes. Consequently, the benefits of fragments programs, i.e. random access to textures, parallelism and memory bandwidth, can be combined with the ability to render large sets of geometric data from graphics memory. In this work, we present an efficient realization of such memory objects on the latest ATI card, the Radeon 9800.

To fully take advantage of this mechanism, efficient GPU realizations of algorithms used to perform particle manipulation are essential. In this paper, we present novel approaches to carry out particle simulation in the fragment units of programmable graphics hardware. At the core of these techniques, we present an optimized sorting routine that is far faster than previous approaches. Built upon this implementation, we have implemented inter-particle collision detection and visibility sorting including hundreds of thousands of primitives. In combination with OpenGL memory objects we present the first particle engine that entirely runs on the GPU *and* includes such effects. This is a significant extension of previous approaches, i.e. [KvdDP03, GRLM03], where the GPU was only exploited to either account for collisions with implicitly defined surfaces or to perform the broad phase in collision detection between polygonal objects.

The remainder of this paper is organized as follows. In the following, we first review related work in the field of GPU programming. In Chapter 3 we introduce the concept of OpenGL SuperBuffers, and we describe their realization on recent ATI cards. The use of parallel fragment units for collision detection and visibility sorting is subject of Chapter 4. Here, we also outline the conceptual design of a particle engine on programmable graphics hardware. In Chapter 5 we show timing statistics for different scenarios including collisions and front-to-back sorting. We conclude the paper with a detailed discussion, and we show further results of our approach.

2. Related Work

To take full advantage of new graphics chip technologies, considerable effort has been spent on the development of algorithms amenable to the intrinsic parallelism and efficient communication on such chips. In many examples, programmable GPUs have been explored to speed up algorithms previously run on the CPU. The computational power and memory bandwidth on such processors has been exploited to accelerate the simulation of both local as well as global illumination effects (e.g. [PDC*03, DS03]). Besides its use for the realistic rendering of geometric data, programmable graphics hardware has also been harnessed for the rendering of volumetric data sets [RSEB*00, EKE01, KPHE02]. The acceleration of image based modeling and rendering techniques on GPUs has been considered for example in [YWB03, LMS03, HMG03]. Recently, a number of researchers have demonstrated the benefits of graphics hardware for the implementation of general techniques of numerical computing [HBSL03, LKHW03, SHN03, MA03, KW03, BFGS03].

The results given in many of these examples show, that for compute bound applications as well as for memory bandwidth bound applications the GPU has the potential to outperform software solutions. However, this statement only holds for such algorithms that can be compiled to a stream program, and which then can be processed by SIMD kernels as provided on recent GPUs. On the other hand, execution speed is not the only concern when mapping algorithms to graphics hardware. Another concern is to avoid any data transfer between the CPU and the GPU. As we have to deal with increasing scene and model complexity, bandwidth requirements become an ever more important issue in a graphics application. By trying to keep both the application program and the rendering module on the graphics subsystem, bandwidth limitations can be avoided thus achieving superior frame rates even when execution speed is not significantly higher.

3. OpenGL SuperBuffers

Our method relies on computing intermediate results on the GPU, saving these results in graphics memory, and then using them again as input to the geometry units to render images in the frame buffer. This process requires application control over the allocation and use of graphics memory; intermediate results are "drawn" into invisible buffers, and these buffers are subsequently used to present vertex data or textures to the GPU.

Our implementation exploits a feature of recent ATI graphics hardware that allows graphics memory to be treated as a render target, a texture, or vertex data. This feature is presented to the application through an extension to OpenGL called *SuperBuffers*. The interface allows the application to allocate graphics memory directly, and to specify how that memory is to be used. This information, in turn, is used by the driver to allocate memory in a format suitable for the requested uses. When the allocated memory is bound to an *attachment point* (a render target, texture, or vertex array), no copying takes place. The net effect for the application program therefore is a separation of raw GPU memory from OpenGLs semantic meaning of the data. Thus, SuperBuffers provide an efficient mechanism for storing GPU computation results and later using those results for subsequent GPU computations.

4. GPU-based Particle Engine

Since we aim at developing a particle engine that entirely runs on the GPU, there is a dire need to reinvent modules of the engine in regard to the particular target architecture. Only then can we expect such a system to achieve significantly better performance rates than a software solution. In the following, we describe the most relevant modules responsible for the state of our system. The state is updated on a per time step basis, where a single time step is comprised of the following events:

- Emission
- Collisionless motion of particles
- Sorting of particles
- Pairing of collision partners
- Collision response
- Enforcement of boundary conditions
- Particle rendering

4.1. Emission

Initially, by any suitable scheme particle positions, $r(t)$, are distributed over the 3D domain, and they are encoded in the RGB components of a server-side 2D RGBA texture map of size $n \times n$. As particles are usually released in different time steps of an animation, each gets assigned a time stamp that is stored in the A-component of this texture. The number of unique time stamps depends on the life time of particles in the animation. As we can only use a fixed number of particles, this life time effectively determines the maximum number of particles to be released per time step. A second texture contains for every particle its current velocity, $v(r,t)$. Initially, this velocity is set to a constant value.

To perform any modifications on particle attributes, a view plane aligned quad covering $n \times n$ fragments is rendered. Both textures are bound to that quad, thus enabling a fragment program to access and to modify these attributes. Uniform attributes for all particles can be specified in additional texture coordinates of the quad. Updated values are simultaneously rendered to separate texture render targets using the `ATI_draw_buffer` extension. In the next pass, these targets become the current containers to be processed.

4.2. Collisionless Particle Motion

In the current implementation, each particle is first streamed by its displacement during time interval dt. The displacement is computed using an Euler scheme to numerically integrate quantities based on Newtonian dynamics:

$$v(r,t+dt) = v(r,t) + v_{ext}(r,t) + \frac{F}{m}dt \tag{1}$$

$$r(t+dt) = r(t) + \frac{1}{2}(v(r,t) + v(r,t+dt))dt \tag{2}$$

Here v_{ext} is an external wind field used to model particle movement, F is an external force, and m is the particle mass. The wind field is stored in a 3D texture, and it can thus be sampled in the fragment program by using particle positions as texture coordinates.

4.3. Sorting

Once collisionless motion of particles has been carried out, the engine performs either of the following tasks: it resolves inter-particle collisions and renders the particles as opaque primitives *or* it sorts particles according to their distance to the viewer and renders the particles as semi-transparent primitives. Both scenarios rely on the sorting of particles. Therefore, a sorting key to be considered in the sort has to be specified.

In the latter scenario, the sorting key is the distances of particles to the viewer. In the former scenario, particle space is assumed to be divided into the cells of a regular lattice of mesh size g_0, and grid cells are uniquely enumerated in ascending z-, y-, x-order. The index of the cell containing a particle is associated with that particle, and it is used further on as sorting key.

4.3.1. Sorting Keys and Identifiers

Sorting keys are computed once at the beginning of the sort. In a single rendering pass, the fragment program either computes the distance of particle positions to the viewer or it calculates the grid cell floating point index by $x/g_0^2 + y/g_0 + z$. In addition, to every particle a unique identifier is associated. Sorting keys and identifiers are rendered into the R- and G-components of a 2D texture render target, which is then sorted in row-major order. After finishing the sort, containers for particle positions and velocities are rearranged according to the arrangement of identifiers.

Care has to be taken when computing identifiers on a chip supporting limited integer or floating point precision. Recent ATI graphics hardware only provides 24 bit internal floating point format. With a 16 bit mantissa and a 7 bit exponent we get $2^{16} = 65536$ distinct values in the range $[2^{h-1} \ldots 2^h]$. As we want to animate far larger particle sets of up to 1 million particles, a different enumeration scheme needs to be developed.

When we let the exponent h take positive values from 1 to 2^4, we get 2^{20} distinct but not equally spaced values. These values can be pre-computed on the CPU and stored in ascending row-major order into a texture map. If we assume 2^{20} particles to be processed per frame, a texture of size $2^{10} \times 2^{10}$ is sufficient to serve as container for these values. To get a unique identifier for each particle, a fragment program simply has to look up the respective identifiers from this container.

At the end of the sorting procedure, from 1D identifiers the initial 2D texture coordinates have to be decoded in order to rearrange particle positions and velocities accordingly. Since we know that values within consecutive sets of

$2^{16}/2^{10}$ rows are equally spaced, for every identifier we only need to determine in which of the $2^4 = 16$ sets it is positioned. This is done by successively subtracting the range $[2^i \dots 2^{i-1}]$ of each set from the identifier until it becomes smaller than zero. Then, the identifier is contained in the ith set, and a final division and modulo operation by the spacing in that set times the number of entries per row yields the appropriate texture address. This address is used to fetch corresponding particle positions and velocities, which are finally output by the fragment program.

4.3.2. Bitonic Sort

To perform the sorting procedure efficiently, we account for the architecture of todays graphics processors. Recent GPUs can be thought of as SIMD computers in which a number of processing units simultaneously execute the same instructions on their own data. GPUs can also be thought of as stream processors, which operate on data streams consisting of an ordered sequence of attributed primitives like vertices or fragments. Considerable effort has been spent on the design of sorting algorithms amenable to the data parallel nature of such architectures. Bitonic sort [Bat68] is one of these algorithms. Bitonic sort first generates a sequence necessary as input for the final merge algorithm. This sequence is composed of two sorted subsequences, where the first is in ascending and the other in descending order. Bitonic merge finally merges both subsequences in order to produce a globally ordered sequence.

Purcell et al. [PDC*03] proposed an implementation of the Bitonic sort on a graphics processor, i.e. the *n*VIDIA GeForceFX graphics accelerator. It was used to sort photons into a spatial data structure, yet providing an efficient search mechanism for GPU-based photon mapping. Comparator stages were entirely realized in a fragment program, including arithmetic, logical and texture operations. The authors report their implementation to be compute limited rather than bandwidth limited, and they achieve a throughput far below the theoretical optimum of the target architecture.

In the following, we present an improved Bitonic sort routine that achieves a performance gain by minimizing both the number of instructions to be executed in the fragment program and the number of texture operations.

4.3.3. The Sorting Pipeline

To explain our implementation, let us put emphasis on some of the characteristics of Bitonic sort when used to sort a 2D texture. As we can see from figure 1, as long as the texture is to be sorted along the rows, in every pass the same instructions are executed for fragments within the same column. Even more precisely, in the kth pass

- the relative address or offset $\triangle r$ of the element that has to be compared is constant
- this offset changes sign every 2^{k-1} columns
- every 2^{k-1} columns the comparison operation changes as well.

The information needed in the comparator stages can thus be specified on a per-vertex basis by rendering column-aligned quad-strips covering $2^k \times n$ pixels. In this way, the output of the geometry units can be directly used as input for the fragment units, thus avoiding most of the numerical computations in the fragment program.

In the kth pass $n/2^k$ quads are rendered, each covering a set of 2^k columns. The constant offset $\triangle r$ is specified as uniform parameter in the fragment program. The sign of this offset, which is either 1 or -1, is issued as varying per-vertex attribute in the first component (r) of one of the texture coordinates. The comparison operation is issued in the second component (s) of that texture coordinate. A less than comparison is indicated by 1, whereas a larger than comparison is indicated by -1. In a second texture coordinate the address of the current element is passed to the fragment program. The fragment program in pseudo code to perform the Bitonic sort finally looks like this:

Row-Wise Bitonic Sort

1 $OP1 = TEX1[r_2, s_2]$
2 $sign = r_1 < 0$? -1 : 1
4 $OP2 = TEX1[r_2 + sign * \triangle r, s_2]$
5 output = $OP1.x * s_1 < OP2.x * s_1$? OP1 : OP2

We should note here that the interpolation of per-vertex attributes during scan-conversation generates a sign value between 1 and -1. Consequently, in line 2 the 1 or -1 values must be reconstructed.

To perform row-wise sorting of texture elements, $\sum_{i=1}^{log(n)} \sum_{j=1}^{i}$ quads have to be generated up front. Each set of quads to be rendered in one pass is stored into a separate display list. Once the texture has been sorted along the rows, another $log(n)$ stages have to be performed. In the ith stage, i additional passes are executed to merge 2^i consecutive rows. Finally, row-wise sorting is performed as described. Since sets of rows are always sorted in ascending order from left to right and from top to bottom, every other set of 2^i rows has to be rearranged to generate a descending sequence required by the Bitonic sort. This operation is implicitly realized in the first pass of each stage.

As one can see from figure 2, the same approach of rendering geometry in order to pass information from the geometry units to the fragment units can be used in the merging of rows as well. The only difference is, that quads now have to be transposed to produce constant values along rows instead of columns.

The final optimization results from the observation that the graphics pipeline as implemented on recent cards is highly optimized for the processing of RGBA samples. As a matter of fact, we pack 2 consecutive entries in each row – including sorting key and identifier – into one single RGBA texture value. This approach is extremely beneficial, because

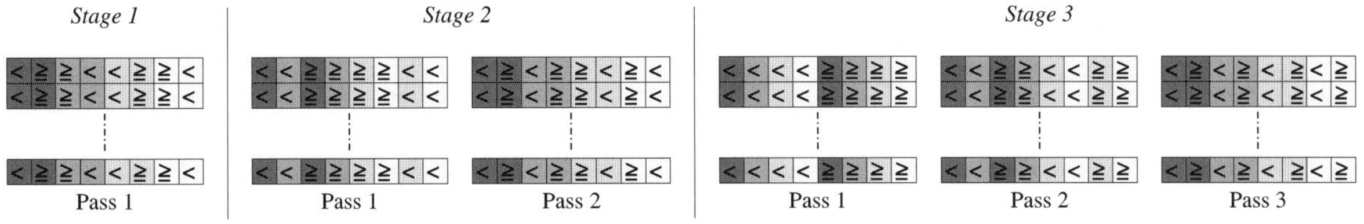

Figure 1: *Bitonic sort of rows of a 2D field. In one row, equal shades of grey indicate elements to be compared. For each element, the operation it has to perform is shown as well.*

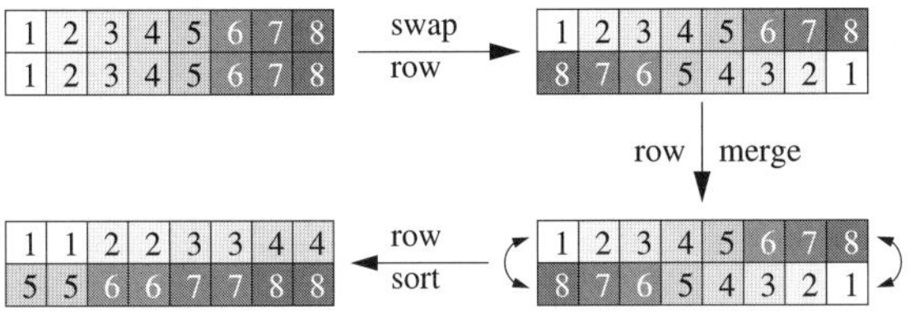

Figure 2: *Bitonic sort of consecutive rows. Every other row is reversed and then merged with its predecessor. Finally, each row is sorted separately, producing a sorted set of elements in every pair of rows.*

it effectively halves the number of fragment operations that have to be performed. In addition, in the first pass of every sorting stage the second operand does not need to be fetched at all. On the other hand, the fragment program only becomes slightly longer because the conditional statement now has to be executed twice. Just at the end of the sorting process is one additional rendering pass required to decode packed texture entries and to rearrange particle positions and velocities according to the arrangement of sorting keys.

4.4. Collision Detection

The collision detection module simultaneously computes for each particle an approximate set of potential collision partners. Only the closest one is kept and used as input for the collision response module. In situations involving multiple collision, resolving these events in parallel can lead to wrong results. Although time-sequential or simultaneous processing of collision events as proposed in [Hah88, Bar89, Mir00] yields correct results, such techniques are not appropriate for the implementation on current graphics architectures.

From the sorted 2D texture, each fragment now fetches the current particle position, and it also fetches a number of particle positions to the left and to the right of this position. Of all these positions the one closest to the current one is kept, and the respective texture coordinate is output to an additional texture render target. Upon completion of collision detection this target is comprised of texture coordinates of the closest potentially colliding partner for every particle.

Although the presented approach provides an effective means for detecting many of the occurred collisions, it has some weaknesses that are worth noting. First, depending on the cell size g_0 many more particles may reside within one cell than can be checked for in the fragment program. Furthermore, because particles are sorted according to cell index only, the closest partner might be the furthest in one row. As a matter of fact, this partner will not be detected. Second, if colliding partners within one cell are arranged in consecutive rows they can not be detected. Third, due to the enumeration of grid cells collisions between particles in adjacent cells can not be detected in general. Fourth, depending on the integration time step and the speed of particles, collisions that occur between successive frames might be missed.

The first drawback can be accounted for by letting the size of grid cells being small enough to only allow for a fixed number of particles in each cell. By considering the same number of potential partners in either row direction, it is guaranteed that collisions within cells are detected.

To overcome the last drawback we essentially sort the set of particles twice. We first build a second set of sorting keys and identifiers from a staggered grid as shown in figure 3. By selecting the closest partner from both grids, only those pairs separated by cell faces in both grids can not be detected. The vast majority of collisions, however, can be determined and resolved.

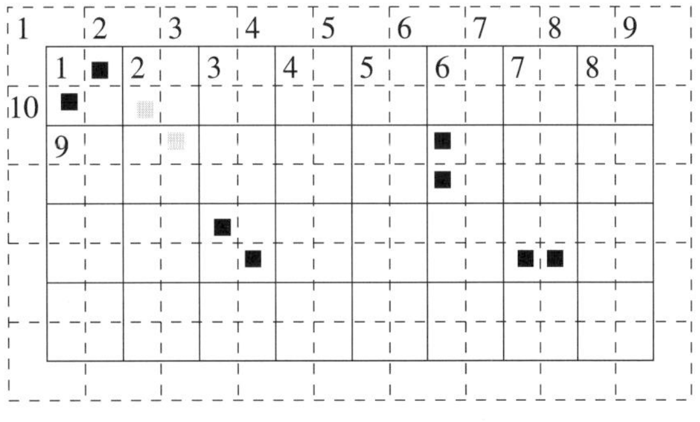

Figure 3: *Initial (dashed) and staggered (solid) grid used for particle enumeration are illustrated in 2D. For the shown particle pairs, collisions can not be detected in the initial grid. Only collisions between greyed-out particles will be missed in the staggered grid.*

4.5. Collision Response

The output of the collision detection module is used as input for the fragment program that simulates collision reactions. Of the potentially colliding partners both position and velocity can be accessed thus providing the information required to compute these reactions. Assuming sphere-like particles with a fixed radius, the fragment program computes the distance between both spheres and tests for a collision. If a collision is ascertained, both particles are backed up in time until the first point of contact is reached.

From conservation of linear momentum and energy, new momentum and thus velocity of the current particle is computed. Spherical rigid bodies can be modeled as well by using additional containers for angular momentum and rotation, and by updating these quantities according to external forces and collision impulses. Updated velocities are used to displace the current particle position according to the time interval that was taken for backtracking. Each fragment finally outputs its position and velocity into the respective textures.

4.6. Boundary Conditions

Once the changes in particle positions and velocities due to inter-particle collisions have been computed, particles are tested for collisions with static parts of the scene. Since direct collision detection between large particle and polygon sets is not feasible in general, we first scan-convert the original scene into a signed distance representation. The distance field is stored in a 3D texture map, thus enabling the fragment program to determine the distance of particles to scene geometry by interpolation in this field.

Although this approach is fairly simple to implement and only needs a single dependent texture operation to access the distance field, for high detailed models it comes with a significant overhead in texture memory. On the other hand, the method is well suited for simulating particle flow over height fields. This scenario only requires a 2D RGBA texture map to store surface normal and height. Once a particle falls below the height field, its position is reset to $p_0 + d_0/(d_0+d_1)*(p_1-p_0)$. In this formula, p_0, p_1 and d_0, d_1 are the previous and the current particle positions, and their distances to the surface, respectively. At the updated point position, the normal h is interpolated and the reflection vector r to model the collision response is computed.

4.7. Rendering

Once particle positions have been updated and saved in graphics memory, these positions are sent through the GPU again to obtain images in the frame buffer. Therefore, updated particle positions are rendered into a vertex array memory object, which is then processed by the geometry engine in a fixed order. The sorting order that has eventually been computed can thus be maintained. In addition to vertex geometry, attributes like color or texture coordinates can be specified in graphics memory as well.

By evaluating the time stamps of particles in a vertex program, the injection of particles can be controlled and particles can be deleted once their life time is expired. Particles that have not yet been born, i.e. their time stamp is larger than the current time step, are discarded by placing these vertices outside the view frustum. Once the particle time stamp modulo the current time step is equal to zero, all particle attributes are reset thus starting a new life cycle.

5. Performance Evaluation and Discussion

To verify the effectiveness of the described particle engine, we investigate its throughput in a variety of different scenarios. All our experiments were run under WindowsXP/Linux on a Pentium 4 2.0 GHz processor equipped with an ATI 9800Pro graphics card.

The proposed stream model for sorting on GPUs achieves a considerable speed up compared to previous approaches. Besides minimizing the number of texture operations, it exploits both the computational power of geometry and fragment processors. By hard-coding information that is required repeatedly in the sorting stages, computational load in the fragment units can be minimized.

Let us now analyze the performance of the sorting routine by sorting differently sized 2D texture maps. To sort a texture of size $n \times n$, $log(n^2) \cdot (log(n^2)+1)/2$ rendering passes are required. In each pass, sorting keys and identifiers of respective operands are accessed, compared and rendered. In the table below, we compare different implementations of the Bitonic sort on recent ATI hardware (including the computation of sorting keys and identifiers as well as the rearrangement of particle containers): *FP* implements the entire sort in a fragment program, *GP/FP* exploits geometry units to pass hard-coded information to the fragment program, and *GP/FP Packed* packs consecutive sorting keys and identifiers in one texture element.

Table 1: Timings (fps) for sorting 2D texture maps.

	128^2	256^2	512^2	1024^2
FP	22	3.5	1	0.2
GP/FP	40	10	3	0.4
GP/FP Packed	148	43	10	2

We see a significant performance gain due to the proposed optimization schemes. Overall, the ultimate implementation computes about 600 MPixels per second, thus almost reaching the theoretical optimum of the graphics card. Even more, by restricting the sorting to only a few sets of consecutive rows we can flexibly select the appropriate frame rate required by the particle engine. Especially in animations where particles are released in time-sequential order this feature enables real-time animations, yet considerably reducing the number of missed collisions. By holding in the same row all particles that are released in one time step, these particles can usually be sorted without sacrificing speed. Although collisions between sets of particles having different time stamps won't be detected, as long as these sets do not mix up entirely, the number of missed collision will be small.

Next, we give timing statistics for the rendering of sets of 256^2, 512^2 and 1024^2 particles. Corresponding screen shots are shown in the color plates below. We analyze the performance of the engine with regard to the simulation of different effects: collisionless particle motion (E1), collisions with a height field (E2), collisionless motion including front-to-back sorting and rendering (E3) and full inter-particle collisions (E4). Inter-particle collision detection includes sorting along texture rows only, and testing a set of 8 particles to the left and to the right of each particle in the fragment program. To compare our system to CPU-based particle engines, (E5) lists the timings for an implementation optimized for CPU processing that uses data-dependent sorting (quick-sort) for front-to-back rendering and is thus equivalent to the GPU experiment (E3).

Table 2: Animation times for large particle sets (fps).

	E1	E2	E3	E4	E5
256^2	640	155	39	133	7
512^2	320	96	8	31	2
1024^2	120	42	1.4	7	0.4

Timings in column E1 essentially show the throughput of the geometry engine combined with OpenGL SuperBuffers. Particles are rendered with associated colors and disabled z-test. A considerable loss in performance can be perceived when using memory objects smaller than 1024^2 – an indication that the graphics card can handle large chunks of data much more efficiently.

As can be seen, even when combining visibility sorting and particle-scene collision detection we are still able to run a real-time animation with about 10 frames per second for a quarter of a million particles. Obviously, animating a million particles puts some load on both the geometry and the fragment subsystem. On the other hand, by restricting the sort to texture rows, we can still perform dynamic simulation of this number of particles with some frames per second.

6. Conclusion

In this paper, we have presented the first particle engine that entirely runs on programmable graphics hardware *and* includes effects such that inter-particle collision and visibility sorting. To fully take advantage of OpenGL memory objects, efficient GPU realizations of algorithms used to perform particle manipulation have been developed. By combining memory objects with floating-point vertex and fragment programs, the system enables real-time animation and rendering of particle dynamics. At run-time, CPU-GPU transfer is completely avoided.

We believe that our work is influential for future research in the field of computer graphics due to several reasons: First, for the first time it has been shown that geometry data can be created, manipulated and rendered on the GPU. By combining vertex and fragment units, it is now possible to simultaneously use the GPU as numerical number cruncher and render server. As this approach allows avoiding any kind of data transfer between the CPU and the GPU, it will significantly speed up applications where numerical computation and rendering of large geometry data is paramount. Second, the particle engine as implemented allows integrating any physical model that requires access to adjacent particles to update particle dynamics. Thus, a variety of grid-less methods to computational simulation of physics based effects can be mapped to graphics hardware. Third, because sorting is at the core of many applications in computer graphics, e.g. occlusion culling, global illumination, unstructured grid rendering, scene graph optimization, the efficient implementation of a sorting algorithm on the render server is extremely beneficial and can be directly used to accelerate a variety of different applications.

References

[Bar89] BARAFF D.: Analytic methods for dynamic simulation of non-penetrating rigid bodies. In *ACM Computer Graphics (Proc. SIGGRAPH '89)* (1989), pp. 223–232.

[Bat68] BATCHER K.: Sorting networks and their applications. In *Proceedings AFIPS 1968* (1968).

[BFGS03] BOLZ J., FARMER I., GRINSPUN E., SCHRÖDER P.: Sparse matrix solvers on the GPU: Conjugate gradients and multigrid. In *ACM Computer Graphics (Proc. SIGGRAPH '03)* (2003), pp. 917–924.

[Boo90] BOON J.: *Lattice Gas Automata: A New Approach to the Simulation of Complex Flows.* Plenum Press, 1990.

[BW97] BARAFF D., WITTKIN A.: Physically based modeling: Principles and practice. ACM Siggraph '97 Course Note, 1997.

[Doo90] DOOLEAN G. (Ed.): *Lattice Gas Methods for Partial Differential Equations.* Addison Wesley Longman, 1990.

[DS03] DACHSBACHER C., STAMMINGER M.:

Translucent shadow maps. In *Proceedings Eurographics Symposium on Rendering 2003* (2003).

[EKE01] ENGEL K., KRAUS M., ERTL T.: High-quality pre-integrated volume rendering using hardware-accelerated pixel shading. In *SIGGRAPH/Eurographics Workshop on Graphics Hardware* (2001).

[GRLM03] GOVINDARAJU N., REDON S., LIN M., MANOCHA D.: Cullide: Interactive collision detection between complex models in large environments using graphics hardware. In *Proceedings ACM SIGGRAPH/Eurographics Conference on Graphics Hardware* (2003).

[Hah88] HAHN J.: Realistic animation of rigid bodies. In *ACM Computer Graphics (Proc. SIGGRAPH '88)* (1988), pp. 173–182.

[HBSL03] HARRIS M., BAXTER W., SCHEUERMANN T., LASTRA A.: Simulation of cloud dynamics on graphics hardware. In *Proceedings ACM SIGGRAPH/Eurographics Workshop on Graphics Hardware* (2003), pp. 12–20.

[HMG03] HILLESLAND K., MOLINOV S., GRZESZCZUK R.: Nonlinear optimization framework for image-based modeling on programmable graphics hardware. In *ACM Computer Graphics (Proc. SIGGRAPH '03)* (2003), pp. 925–934.

[KPHE02] KNISS J., PREMOZE S., HANSEN C., EBERT D.: Interactive translucent volume rendering and procedural modeling. In *Proceedings IEEE Visualization* (2002).

[KvdDP03] KNOTT D., VAN DEN DOEL K., PAI D. K.: Particle system collision detection using graphics hardware. In *SIGGRAPH 2003 Sketch* (2003).

[KW03] KRUEGER J., WESTERMANN R.: Linear algebra operators for GPU implementation of numerical algorithms. In *ACM Computer Graphics (Proc. SIGGRAPH '03)* (2003), pp. 908–916.

[LKHW03] LEFOHN A., KNISS J., HANSEN C., WHITAKER R.: Interactive deformation and visualization of level set surfaces using graphics hardware. In *Procceedings IEEE Visualization* (2003).

[LMS03] LI M., MAGNOR M., SEIDEL H.-P.: Hardware-accelerated visual hull reconstruction and rendering. In *Proceedings of Graphics Interface* (2003), pp. 12–20.

[LT93] LEECH J., TAYLOR R.: Interactive modeling using particle systems. In *Proc. 2nd Conference on Discrete Element Methods* (1993).

[MA03] MORELAND K., ANGEL E.: The FFT on a GPU. *Proceedings ACM SIGGRAPH/Eurographics Workshop on Graphics Hardware* (2003), 112–119.

[McA00] MCALLISTER D.: The design of an api for particle systems, 2000.

[MGAK03] MARK W., GLANVILLE R., AKELEY K., KILGARD M.: Cg: A system for programming graphics hardware in a C-like language. In *ACM Computer Graphics (Proc. SIGGRAPH '03)* (2003), pp. 896–907.

[Mic02] MICROSOFT: DirectX9 SDK. http://www.microsoft.com/ DirectX, 2002.

[Mir00] MIRTICH B.: Timewarp rigid body simulation. In *ACM Computer Graphics (Proc. SIGGRAPH '00)* (2000), pp. 193–200.

[Mon88] MONAGHAN J.: An Introduction to SPH. *cpc 48* (1988), 89–96.

[PDC*03] PURCELL T., DONNER C., CAMMARANO M., JENSEN H., HANRAHAN P.: Photon mapping on programmable graphics hardware. In *Proceedings ACM SIGGRAPH/Eurographics Workshop on Graphics Hardware* (2003), pp. 41–50.

[Ree83] REEVES T.: Particle systems - a technique for modelling a class of fuzzy objects. *ACM Computer Graphics (Proc. SIGGRAPH '83)* (1983).

[RSEB*00] REZK-SALAMA C., ENGEL K., BAUER M., GREINER G., T. E.: Interactive volume rendering on standard pc graphics hardware using multi-textures and multi-stage rasterization. In *Eurographics Workshop on Graphics Hardware* (2000), pp. 109–119.

[SHN03] SHERBONDY A., HOUSTON M., NAPEL S.: Fast volume segmentation with simultaneous visualization using programmable graphics hardware. In *Procceedings IEEE Visualization* (2003).

[Sim90] SIMS K.: Particle animation and rendering using data parallel computation. In *Computer Graphics (Siggraph '90 proceedings)* (1990), pp. 405–413.

[YWB03] YANG R., WELCH G., BISHOP G.: Real-time consensus-based scene reconstruction using commodity graphics hardware. In *Proceedings of Pacific Graphics* (2003), pp. 23–31.

Graphics Hardware (2004)
T. Akenine-Möller, M. McCool (Editors)

Hardware-based Simulation and Collision Detection for Large Particle Systems

A. Kolb* and L. Latta† and C. Rezk-Salama*

*Computer Graphics and Multimedia Systems Group, University of Siegen, Germany
†2L Digital, Mannheim, Germany

Abstract

Particle systems have long been recognized as an essential building block for detail-rich and lively visual environments. Current implementations can handle up to 10,000 particles in real-time simulations and are mostly limited by the transfer of particle data from the main processor to the graphics hardware (GPU) for rendering.
This paper introduces a full GPU implementation using fragment shaders of both the simulation and rendering of a dynamically-growing particle system. Such an implementation can render up to 1 million particles in real-time on recent hardware. The massively parallel simulation handles collision detection and reaction of particles with objects for arbitrary shape. The collision detection is based on depth maps that represent the outer shape of an object. The depth maps store distance values and normal vectors for collision reaction. Using a special texture-based indexing technique to represent normal vectors, standard 8-bit textures can be used to describe the complete depth map data. Alternately, several depth maps can be stored in one floating point texture.
In addition, a GPU-based parallel sorting algorithm is introduced that can be used to perform a depth sorting of the particles for correct alpha blending.

Categories and Subject Descriptors (according to ACM CCS): I.3.1 [Computer Graphics]: Graphics processors I.3.5 [Computer Graphics]: Boundary representations I.3.7 [Computer Graphics]: Animation

1. Introduction

Physically correct particle systems (PS) are designed to add essential properties to the virtual world. Over the last decades they have been established as a valuable technique for a variety of applications, e.g. deformable objects like cloth [VSC01] and volumetric effects [Har03].

Dynamic PS have been introduced by [Ree83] in the context of the motion picture Star Trek II. Reeves describes basic motion operations and basic data representing a particle - both have not been altered much since. An implementation on parallel processors of a super computer has been done by [Sim90]. [Sim90] and [McA00] describe many of the velocity and position operations of the motion simulation also used in our PS.

Real-time PS are often limited by the fill rate or the CPU to graphics hardware (GPU) communication. The fill rate is often a limiting factor when there is a high overdraw due to relatively large particle geometries. Using a large number of smaller particles decreases the overdraw and the fill rate limitation looses importance. The bandwidth limitation now dominates the system. Sharing the graphics bus with many other rendering tasks allows CPU-based PS to achieve only up to 10,000 particles per frame in typical real-time applications. A much larger number of particles can be used by minimizing the amount of communication of particle data by integrating simulation and rendering on the GPU.

Stateless PS, i.e. all particle data can be computed by closed form functions based on a set of start values and the current time, have been implemented using vertex shaders (cf. [NVI03]). However, state-preserving PS can utilize numerical, iterative integration methods to compute the particle data from previous values and a dynamically changing environment. They can be used in a much wider range of applications.

While collision reaction for particles is a fairly simple application of Newtonian physics, collision detection can be a rather complex task w.r.t. the geometric representation of the collider object. [LG98] gives a good overview on collision detection techniques for models represented as CSG, polygonal, parametric or implicit surfaces. There are three basic *image-based* hardware accelerated approaches to collision detection based on depth buffers, stencil buffers or occlusion culling. However, all these techniques use the GPU to generate spatial information which has to be read back from the GPU to the CPU for further collision processing.

The technique presented in this paper uses the "stream processing" paradigm, e.g. [PBMH02], to implement PS simulation, collision detection and rendering completely on the fragment processor. Thus a large number of particles can be simulated using the state-preserving approach. The collision detection is based on an implicit, image based object boundary representation using a sequence of depth maps similar to [KJ01]. Several approaches are presented to store one, two or six depth maps in a single texture. Storing the normal vectors for collision reaction is realized using a texture-based normal indexing technique.

The remainder of this paper is organized as follows. Section 2 gives an overview over works related to this paper. The GPU based simulation for large particle systems is described in section 3. Collision detection on the GPU is discussed in section 4. Results and conclusions are given in sections 5 and 6 respectively.

2. Prior work

This section describes prior works related to particle systems (PS) and their implementation using graphics hardware (section 2.1). Additionally, we briefly discuss collision detection approaches (section 2.2), techniques to generate implicit representations for polygonal objects (section 2.3) and the compression of normal vectors (section 2.4).

2.1. Stateless particle systems

Some PS have been implemented with vertex shaders on programmable GPUs [NVI03]. However, these PS are stateless, e.g. they do not store the current positions of the particles. To determine a particle's position a closed form function for computing the current position only from initial values and the current time is needed. As a consequence such PS can hardly react to a dynamically changing environment.

Particle attributes besides velocity and position, e.g. the particle's orientation, size and texture coordinates, have generally much simpler computation rules, e.g. they might be calculated from a start value and a constant factor of change over time.

So far there have been no state-preserving particle systems fully implemented on the GPU.

2.2. Collision detection techniques

The field of collision detection is one of the most active in recent years. Lin and Gottschalk [LG98] give a good overview on various collision detection techniques and a wide range of applications, e.g. game development, virtual environments, robotics and engineering simulation.

There are three basic hardware accelerated approaches based on depth buffers, stencil buffers and occlusion culling. All approaches are image based and thus their accuracy is limited due to the discrete geometry representation.

Stencil buffer and depth buffer based approaches like [BW03, HZLM02, KOLM02] use the graphics hardware to generate proximity, collision or penetration information. This data has to be read back to the CPU to perform collision detection and reaction. Usually, these techniques use the graphics hardware to detect pairs of objects which are potentially colliding. This process may be organized hierarchically to get either more detailed information or to reduce the potentially colliding objects on a coarser scale.

Govindaraju et.al. [GRLM03] utilize hardware accelerated occlusion queries. This minimizes the bandwidth needed for the read-back from the GPU. Again, the collision reaction is computed on the CPU.

2.3. Implicit representation of polygonal objects

Implicit representations of polygonal objects have advantages in the context of collision detection, since the distance of any 3D-point is directly given by the value of the implicit model representation, the so-called *distance-map*.

Nooruddin and Turk [NT99, NT03] introduced a technique to convert a polygonal model in an implicit one using a scanline conversion algorithm. They use the implicit representation to modify the object with 3D morphological operators.

Kolb and John [KJ01] build upon Nooruddin and Turk's algorithm using graphics hardware. They remove mesh artifacts like holes and gaps or visually unimportant portions of objects like nested or overlapping parts. This technique generates an approximate distance map of a polygonal model, which is exact on the objects surface w.r.t. to the visibility of object points and the resolution of the depth buffer.

2.4. Compression of normal vectors

For collision reaction, an efficient way to store an object's normal, i.e. the collision normal, at a particular point on the object's surface is needed. Deering [Dee95] notes, that angular differences below 0.01 radians, yielding some $100k$ normal vectors, are not visually recognizable in rendering. Deering introduces a normal encoding technique which requires several trigonometric function calls per normal.

In our context we need a normal representation technique

which is space and time efficient. Possible decoding of the vector components must be as cheap as possible, while the encoded data must be efficiently stored in textures.

Normal maps store normal vectors explicitly and are not space efficient. Applying an optional DXT-compression results in severe quantization artifacts (cf. [ATI03]).

Sphere maps, cube maps and parabolic maps, commonly used to represent environmental information, may be used to store normal vectors. Sphere maps heavily depend on a specific viewing direction. Cube maps build upon a 3D-index, i.e. a point position in 3-space. Parabolic maps need two textures to represent a whole sphere. Additionally, they only utilize the inscribed circle of the texture.

2.5. Other related works

Green [Gre03] describes a cloth simulation using simple grid-aligned particle physics, but does not discuss generic particle systems' problems, like allocation, rendering and sorting of PS. The photon mapping algorithm described by Purcell et.al. [PDC*03] uses a sorting algorithm similar to the odd-even merge sort presented in section 3.3.3. However, their algorithm does not show the necessary properties to exploit the high frame-to-frame coherence of the particle system simulation.

3. Particle simulation on Graphics Hardware

The following sections describe the algorithm of a state-preserving particle system on a GPU in detail. After a brief overview (section 3.1), the storage (section 3.2) and then the processing of particles is described (section 3.3).

3.1. Algorithm Overview

The particle simulation consists of six basic steps:

1. Process birth and death
2. Update velocities
3. Update positions
4. Sort for alpha blending (optional)
5. Transfer texture data to vertex data
6. Render particles

The state-preserving particle system stores the velocities and positions (step 2. and 3.) of all particles in textures, which are also render targets. In one rendering pass the texture with particle velocities is updated, performing a single time step of an iterative integration. Here acceleration forces and collision reactions are applied. A second rendering pass updates the position textures in a similar way, using the velocity texture. Depending on the integration method it is possible to skip the velocity update pass, and directly integrate the position from accelerations (cf. section 3.3.2).

Optionally, in step 4. the particle positions can be sorted depending on the viewer distance to avoid rendering artifacts. The sorting performs several additional rendering passes on textures that contain the particle-viewer distance and a reference to the particle itself.

Then the particle positions are transferred from the position texture to a vertex buffer and the particles are rendered as point sprites, triangles or quads.

3.2. Particle data storage

The positions and velocities of all active particles are stored in floating point textures using the three color components as x, y and z coordinates. The texture itself is also a render target, so it can be updated with the computed positions and velocities. Since a texture cannot be used as input and output at the same time, we use a pair of these textures and a double buffering technique (cf. figure 1). Depending on the integration algorithm the explicit storage of the velocity texture can be omitted (cf. section 3.3.2).

Other particle attributes like mass, orientation, size, color, and opacity are typically static or can be computed using a simple stateless approach (cf. section 2.1). To minimize the upload of static attribute parameters we introduce particle types. Thus the simulation of these attributes uses one further texture to store the time of birth and a reference to the particle type for each particle (cf. figure 1). To model more complex attribute behavior, simple key-frame interpolation over the age of the particle can be applied.

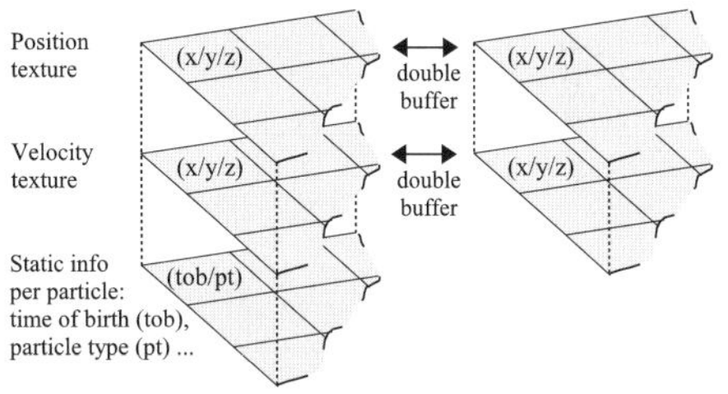

Figure 1: *Data storage using double buffered textures*

3.3. Simulation and rendering algorithm

3.3.1. Process birth and death

Assuming a varying number of short-living particles, the particle system must be able to process the birth of a new particle (*allocation*) and the death of a particle (*deallocation*).

Since allocation problems are serial by nature, this cannot be done efficiently with a data-parallel algorithm on the GPU. Therefore an available particle index is determined on the CPU using either a stack filled with all available indices or a heap data structure that is optimized to always return the smallest available index. The heap data structure guarantees that particles in use remain packed in the first portion of the textures. The following simulation and rendering steps only need to update that portion of data, then. The initial particle data is determined on the CPU and can use complex algorithms, e.g. probability distributions (cf. McAllister [McA00]).

A particle's death is processed independently on the CPU and GPU. The CPU registers the death of a particle and adds the freed index to the allocator. The GPU does an extra pass to determine the death of a particle by the time of birth and the computed age. The dead particle's position is simply moved to invisible areas. As particles at the end of their lifetime usually fade out or fall out of visible areas anyway, the extra clean-up pass rarely needs to be done.

3.3.2. Update velocities and position

Updating a particle's velocity and position is based on the Newtonian laws of motion. The actual simulation is implemented in a fragment shader. The shader is executed for each pixel of the render target by rendering a screen-sized quad. The double buffer textures are alternately used as render target and as input data stream, containing the velocities and positions from the previous time step.

There are several operations that can be used to manipulate the velocity (cf. [Sim90] and [McA00] for more details): global forces (e.g. gravity, wind), local forces (attraction, repulsion), velocity dampening, and collision responses. For our GPU-based particle system these operations need to be parameterized via fragment shader constants.

A complex local force can be applied by mapping the particle position into a 2D or 3D texture containing flow velocity vectors $\vec{\mathbf{v}}_{\text{flow}}$. Stoke's law is used to derive a dragging force:

$$\vec{\mathbf{F}}_{\text{flow}} = 6\pi\eta\, r(\vec{\mathbf{v}}_{i-1} - \vec{\mathbf{v}}_{\text{flow}})$$

where η is the flow viscosity, r the particle radius and $\vec{\mathbf{v}}_{i-1}$ the particle's previous velocity.

The new velocity $\vec{\mathbf{v}}_i$ and position $\mathbf{P}$ is derived from the accumulated global and local forces $\vec{\mathbf{F}}$ using simple Euler integration.

Alternatively, Verlet integration (cf. [Ver67]) can be used to avoid the explicit storage of the velocity by utilizing the position information $\mathbf{P}_{i-2}$. The great advantage is that this technique reduces memory consumption and removes the velocity update rendering pass.

Verlet integration uses a position update rule based only on the acceleration:

$$\mathbf{P}_i = 2\mathbf{P}_{i-1} - \mathbf{P}_{i-2} + \vec{\mathbf{a}}\Delta_i^2$$

Using Euler integration, collision reaction is based on a change of the particle's velocity. Splitting the velocity vector $\vec{\mathbf{v}}_i$ into a normal component $\vec{\mathbf{v}}_i^{\perp}$ and a tangential component $\vec{\mathbf{v}}_i^{\parallel}$ the velocity after the collision can be computed applying friction ν and resilience ε:

$$\vec{\mathbf{v}}_i = (1-\nu)\vec{\mathbf{v}}_i^{\parallel} - \varepsilon\vec{\mathbf{v}}_i^{\perp}$$

To avoid velocities from slowing down close to zero, the friction slow-down should not be applied if the overall velocity is smaller than a given threshold.

Having collider with sharp edges, e.g. a height field, or two colliders close to each other, the collision reaction might push particles into a collider. In this case a caught particle ought to be pushed out in the next simulation step.

To handle this situation, the collision detection is done twice, once with the previous and once with the expected position $\mathbf{P}_i^*$ based on velocity $\vec{\mathbf{v}}_i^*$. This allows differentiating between particles that are about to collide and those having already penetrated (cf. figure 2). The latter must be pushed in direction of the shortest way out of the collider. This direction can be guessed from the normal component of the velocity:

$$\vec{\mathbf{v}}_i = \begin{cases} \vec{\mathbf{v}}_i^* & \text{if } \left(\vec{\mathbf{v}}_i^* \cdot \hat{\mathbf{n}}\right) \geq 0 \quad \text{(heading outside)} \\ \vec{\mathbf{v}}_i^{\perp} - \vec{\mathbf{v}}_i{}^* & \text{if } \left(\vec{\mathbf{v}}_i^* \cdot \hat{\mathbf{n}}\right) < 0 \quad \text{(heading inside)} \end{cases}$$

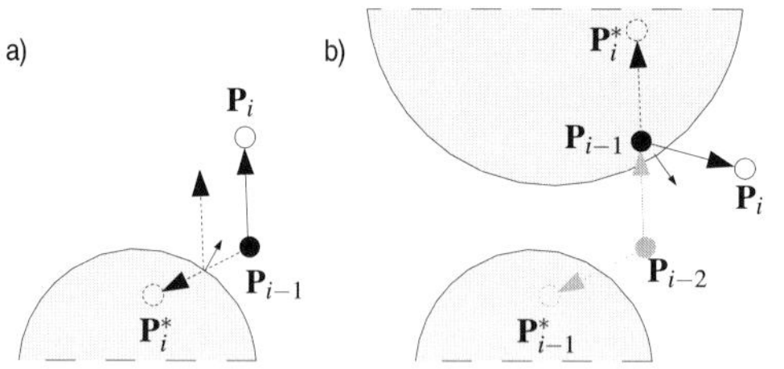

Figure 2: *Particle collision: a) Reaction before penetration; b) Double collision with caught particle and push-back.*

Verlet integration cannot directly handle collision reactions in the way discussed above. Here position manipulations are required to implicitly change the velocity in the following frames.

3.3.3. Sorting

If particles are blended using a non-commutative blending mode, a depth-based sorting should be applied to avoid artifacts.

A particle system on the GPU can be sorted quite efficiently with the parallel sorting algorithm "odd-even merge sort" (cf. Batcher [Bat68]). Its runtime complexity is independent of the data's sortedness. Thus, a check whether all data is already in sequence does not need to be performed on the GPU, which would be rather inefficient. Additionally, with each iteration the sortedness never decreases. Thus, using the high frame-to-frame coherence of the particle order, the whole sorting sequence can be distributed over 20 - 50 frames. This, of course, leads to an approximate depth sortedness, which, in our examples, does not yield any visual artifacts.

The basic principle of odd-even merge sort is to divide the data into two halves, to sort these and then to merge the two halves. The algorithm is commonly written recursively, but a closer look at the resulting sorting network reveals its parallel nature. Figure 3 shows the sorting network for eight values. Several consecutive comparisons are independent of each other and can be grouped for parallel execution (vertical lines in figure 3).

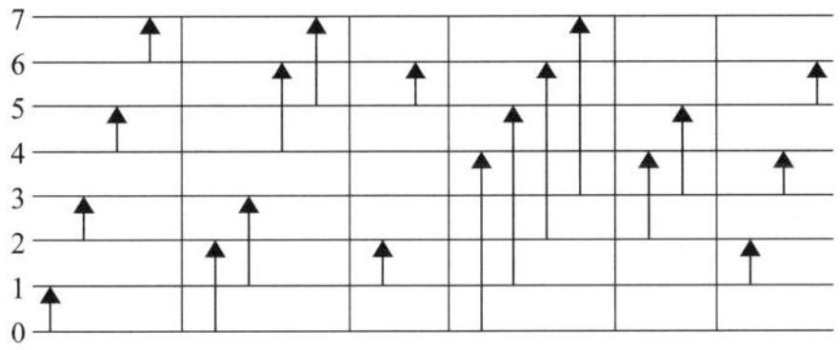

Figure 3: *Odd-Even sorting network for eight values; arrows mark comparison pairs.*

The sorting requires $\frac{1}{2}\log_2^2 n + \frac{1}{2}\log_2 n$ passes, where n is the number of elements to sort. For a 1024×1024 texture this leads to 210 rendering passes. Running all 210 passes each frame is far too expensive, but spreading the whole sorting sequence over 50 frames, i.e. 1 - 2 seconds, reduces the workload to $4-5$ passes each frame.

The sorting algorithm requires an additional texture containing the particle-viewer distance. The distance in this texture is updated after the position simulation. After sorting the rendering step looks up the particle attributes via the index in this texture.

3.3.4. Render particles

The copying of position data from a texture to vertex data is an upcoming hardware feature in PC GPUs. DirectX and OpenGL offer the vertex textures technique (vertex shader 3.0 rsp. `ARB_vertex_shader` extension). Unfortunately there is no hardware supporting this feature at the moment.

Alternatively "über-buffers" (also called super buffers; cf. [Per03]) can be used. This functionality is already available in current GPUs, but up to now it is only supported by the OpenGL API. The current implementation uses the vendor specific `NV_pixel_data_range` extension (cf. [NVI04]).

The transferred vertex positions are used to render primitives to the frame buffer. In order to reduce the workload of the vertex unit, particles are currently rendered as point sprites instead of as triangles or quads. The disadvantage though is that particles are always axis-aligned. To allow a 2D-rotation, texture coordinates are transformed in the fragment shader.

4. Collision detection

In this section we describe the implicit object representation used for collision detection (section 4.1). Furthermore, the normal indexing technique (section 4.2) and various approaches to represent depth maps in textures are introduced (section 4.3).

4.1. Implicit model representation

We use an image-based technique similar to [KJ01] to represent an objects outer boundary by a set of depth maps. These depth maps contain the distance to the object's boundary and the surface normal at the relevant object point.

Each depth map DM_i, $i = 1,\dots,k$ stores the following information:

1. distance $\text{dist}(x,y)$ to the collider object in projection direction for each pixel (x,y)
2. normal vector $\hat{\mathbf{n}}(x,y)$ at the relevant object surface point
3. transformation matrix $T_{OC\to DC}$ mapping from collider object coordinates to depth map coordinates, i.e. the coordinate system in which the projection was performed
4. z_{scale} scaling value in z-direction to compensate for possible scaling performed by $T_{OC\to DC}$

The object's interior is assigned with negative distance values. Assuming we look from outside onto the object and orthographic projection is used, the distance value $f(\mathbf{P})$ for a point $\mathbf{P}$ is computed using the transformation $T_{OC\to DC}$:

$$f(\mathbf{P}) = z_{scale} \cdot \left(\text{dist}(p'_x, p'_y) - p'_z\right), \qquad (1)$$
$$\text{where } \mathbf{P}' = (p'_x, p'_y, p'_z)^T = T_{OC\to DC}\mathbf{P}$$

$T_{OC\to DC}$ usually also contains the transformation to texture coordinates. Thus fetching the depth value $\text{dist}(p'_x, p'_y)$ is a simple texture lookup at coordinates (p'_x, p'_y).

Taking several depth maps into account, the most appropriate depth for point $\mathbf{P}$ has to be determined. The following definition guarantees that $\mathbf{P}$ is outside of the collider if at least one depth map has recognized $\mathbf{P}$ to be exterior:

$$f(\mathbf{P}) = \begin{cases} \max\{f_i(\mathbf{P})\} & \text{if } f_i(\mathbf{P}) < 0\ \forall i \\ \min\{f_i(\mathbf{P})\ :\ f_i(\mathbf{P}) > 0\} & \text{else} \end{cases}$$

where $f_i(\mathbf{P})$ denotes the signed distance value w.r.t. depth map DM_i.

Handling several depth maps DM_i, $i = 1,\dots,k$, $f(\mathbf{P})$ can be computed iteratively:

$$\vee \left.\begin{matrix} (f(\mathbf{P}) < 0 \wedge f_i(\mathbf{P}) > f(\mathbf{P})) \\ (f_i(\mathbf{P}) > 0 \wedge f_i(\mathbf{P}) < f(\mathbf{P})) \end{matrix}\right\} \Rightarrow (f(\mathbf{P}) \leftarrow f_i(\mathbf{P})) \qquad (2)$$

where $f(\mathbf{P})$ is initially set to a large negative value, i.e. $\mathbf{P}$ is placed "far inside" the collider.

If $\mathbf{P}' = T_{OC \to DC}\mathbf{P}$ lies outside the view volume for the current depth map or the texture lookup $\text{dist}(p'_x, p'_y)$ results in the initial background value, e.g. no distance to the object can be computed, this value may be ignored as *invalid vote*. Alternatively, if the view volume encloses the complete collider object, the invalid votes can be declared as "far outside". To avoid erroneous data due to clamping of (p'_x, p'_y), we simply keep a one pixel border in the depth map unchanged with background values to indicate an invalid vote.

A fragment shader computes the distance using rule (2) and applies a collision reaction when the final $f(\mathbf{P})$ distance is negative.

Note, that this approach may have problems with small object details in case of an insufficient buffer resolution. Problems can also be caused by local surface concavities. In many cases, these local concavities can be reconstructed with properly placed depth map view volumes (cf. section 5).

4.2. Quantization and indexing of normal vectors

Explicitly storing normal vectors using $8, 16$ or 32 bit per component is sufficient within certain error bound (cf. [ATI03]). Since we store unit vectors, most of the used 3-space remains unused, though. The depth map representation requires a technique which allows the retrieval of normal vectors using a space- and time efficient indexing method. Indices must be encoded in depth maps and the reconstruction of the normal vector in the fragment shader must be efficient.

The normal representation, which is implemented by indexing into a normal index texture, should have the following properties:

1. the complete coordinate space $[0,1]^2$ of a single texture is used
2. decoding and ideally encoding is time efficient
3. sampling of the directional space is as regular as possible

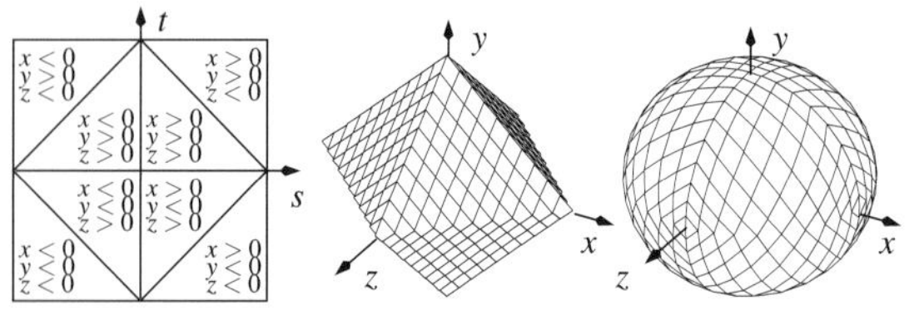

Figure 4: *The eight octants of the L_1-parametrization (left), the octahedron after applying l_1 (middle) and the sampling of the unit sphere (right).*

Cube maps can not be used, since the index to look-up the function value is already a vector with three components. Sphere maps, commonly used as reflection maps, heavily depend on a specific direction, e.g. the viewing direction for the reflection computation. On the other hand, parabolic maps show a very uniform parameterization of the hemi-sphere (cf. Heidrich and Seidel [HS98]), but two of these textures are needed to span the whole sphere.

We propose the following mapping, which is based on the L_1-*norm*: $\|\vec{\mathbf{v}}\|_1 = |v_x| + |v_z| + |v_z|$:

$$l_1(s,t) = \begin{cases} \begin{pmatrix} s \\ t \\ 1-|s|-|t| \end{pmatrix} & \text{if } |s|+|t| \le 1 \\ \begin{pmatrix} \text{sgn}(s)(1-|t|) \\ \text{sgn}(t)(1-|s|) \\ 1-|s|-|t| \end{pmatrix} & \text{if } |s|+|t| > 1 \end{cases} \tag{3}$$

where $s,t \in [-1,1]$. l_1 maps $(s,t) \in [-1,1]^2$ onto the L_1-unit sphere, i.e. the unit octahedron (cf. figure 4). Applying an additional affine transformation, we get a continuous mapping of the standard texture-space $(s,t) \in [0,1]^2$ onto the octahedron. The resolution of the texture-space naturally implies the resolution of the sphere, i.e. the normal space.

It should be pointed out, that the L_1-parametrization proposed above can easily be used to represent any directional data, e.g. reflection maps.

4.3. Depth map representation

Ideally, we want to encode as many depth values and normal vectors as possible into a single texture, thus keeping the amount of data to be transfered and kept in the graphics hardware memory as small as possible.

Throughout our experiments, we have investigated the following depth map formats:

Floating point depth map
The simplest, but most storage ineffient variant uses a floating point texture to store the surface normals uncompressed in the R, G, B-components and the alpha-channel to hold the distance value.

8-bit fixed point depth map
This variant uses a standard *RGBA*-texture with 8 bit per component. Here the R, G-components contain the index into the normal texture, whereas G, A store the depth value, thus having a depth-resolution of 16-bit fixed point. The normal index texture with resolution 256×256 is build using the L_1-parametrization technique described in section 4.2. The *RGB*-components of this texture store the normal vectors, which are looked up using the index stored in the depth map.

16-bit floating point depth map (front-back)
Combining orthographic projection with depth compare function `LESS` generates a *front* depth map. Naturally the depth map taking the inverse z-direction (depth compare

Figure 5: *Sample applications: "bunny in the snow" (left) and "Venus-fountain" (middle and right)*

function GREATER) is a usefull *back* depth map counterpart. We can easily represent both of these depth maps in one 16-bit texture. Here the R,G components store the normal texture indices, where two 8 bit indices are packed into a single 16-bit float. The B,A components store the depth value for the front and back map respectively.

8-bit fixed point cubic depth maps ("depth cube")
Another variant is to use cube maps to represent depth maps. In this case perspective projection w.r.t. the view volume center is applied. The depth map representation is analog to the 8-bit fixed point variant.

Generally, the different types of depth maps can be combined for collision detection within a single fragment shader, in order to utilize the advantages of the various types for the specific collider object (cf. section 5.2).

The depth cube variant uses perspective projection, whereas the other variants use orthographic projection only. Using perspective projection during the depth map generation distorts the z-values. To avoid this, the vertices of the collider object are passed w.r.t. the normalized view volume $[-1,1]^3$ to the fragment shader. The shader simply uses this information to compute the depth values $\text{dist}(x,y)$ relative to the center of the view volume.

To compute the distance value for a point $\mathbf{P} \in \mathbb{R}^3$ w.r.t. the depth cube, the transformation $T_{OC \to DC}$ in depth map coordinates (cf. section 4.1) does not contain the perspective projection. $T_{OC \to DC}$ transforms into the normalized view volume $[-1,1]^3$ only, thus picking the corresponding depth value is just a cube map texture lookup

$$\text{dist}(p'_x, p'_y, p'_z), \ \mathbf{P}' = T_{OC \to DC}\mathbf{P}$$

The distance value for a point $\mathbf{P} \in \mathbb{R}^3$ is computed as

$$f(\mathbf{P}) = \left(1 - \frac{\text{dist}(p'_x, p'_y, p'_z)}{\|\mathbf{P}'\|}\right) \left\| \begin{pmatrix} s_x \cdot p'_x \\ s_y \cdot p'_y \\ s_z \cdot p'_z \end{pmatrix} \right\|, \ \mathbf{P}' = T_{OC \to DC}\mathbf{P}$$

where s_x, s_y, s_z are the scaling factors from the normalized view volume $[-1,1]^3$ to the view volume's extends in collider object coordinates.

The placement of the depth cube w.r.t. the collider object specifies the depth compare function for the generation of the depth maps. In the default situation, where the view volume's center is outside the object, the depth compare function LESS is used. Otherwise, if the center is inside the collider object, GREATER is applied and the fragment shader which computes the distance has to negate the distance value (cf. equation 1).

5. Results

Several tests have been made with different setups for the particle system, e.g. number of particles, sorting, complexity of collider etc. We discuss the general particle system (PS) issues in section 5.1 and describe different aspects on collision dection in section 5.2. Section 5.3 gives some hardware aspects.

The presented particle system was implemented in *Cg* and tested on *NVIDIA Geforce FX 5900 XT* graphics hardware, which represents the first generation of floating-point GPUs.

5.1. Particle simulation

Using a position texture of size 1024×1024, our PS is capable of simultaneously rendering and updating a maximum of 1 million particles at about 10 frames per second. This implementation uses Euler integration, point sprites for rendering, no depth sorting and no collision detection.

In a typical application a particle texture of size 512×512 can be rendered in real-time (about 15 fps) including depth sorting (5 passes per frame) and collision detection (one depth cube). Performance measurement for the fully-featured collision detection is given in the next section. Following the clear trend towards increasing parallelism, a significant performance enhancement is expected with the forthcoming second generation of floating-point GPUs.

Figure 5 shows two sample applications using a quarter million particles. In example "bunny in the snow" each particle is rendered as a snow flake, i.e. the particles velocity is set to zero after a collision has been detected. The collision

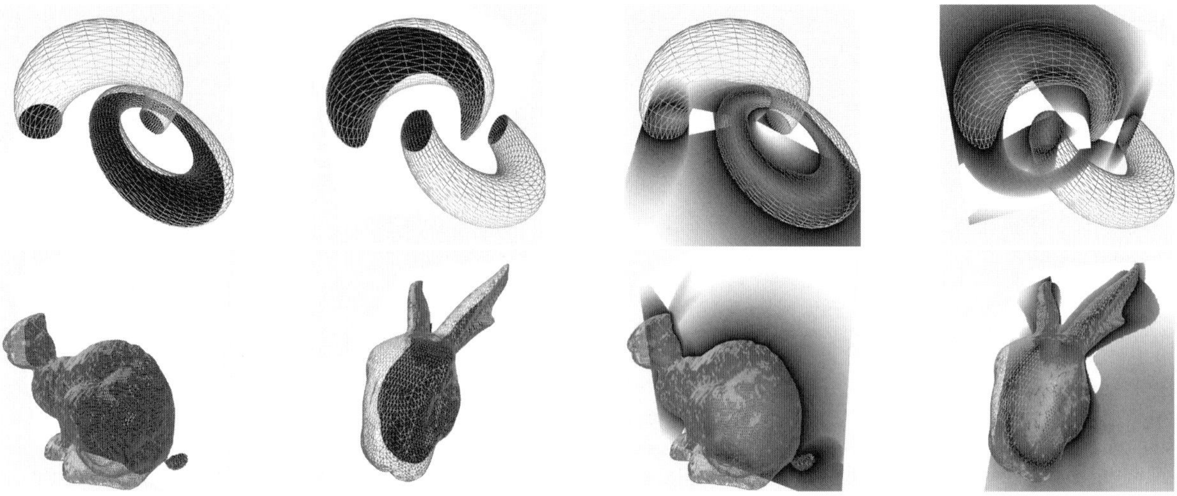

Figure 6: *Visualization of the implicit torii model (top row) and the implicit bunny model (bottom row) along slices: Interior/exterior classification (left) and approximate distance map plotting absolute distance values (right). Additionally, the wireframe model is rendered.*

detection uses one depth cube and one 16-bit front-back texture. The simulation uses depth sorting and runs with 15 fps. The second example, the "Venus fountain", also simulates 512^2 particles. The implicit object boundary is represented using three 16-bit front-back textures and one 8-bit fixed point texture. This examples runs with 10 fps.

5.2. Depth map based collision detection

The main difference between the above presented depth map formats lies in the number of depth maps used and in the projection type. In situations, where collisions occur only from one direction and the collider object has a rather simple shape, a single 8-bit fixed point depth map may result in a proper interior/exterior classification. If there is no restriction to the potential collision direction, the complete collider object has to be reconstructed. Here, either one depth cube or three orthogonal 16-bit front-back textures are used. Concave regions may have to be treated using additional depth maps.

Concerning distance values, depth cubes work well for sphere-like collider objects (cf. figure 6). If the model has many concavities, is strongly twisted or is partly nested, the reconstruction of the distance values based on depth maps leads only to coarse approximations.

In our experiments we use six to 15 depth maps to represent the collider object boundary without restriction to the potential collision direction. Testing a quarter million particles for collision takes 7, 9 and 12 ms using the 8-bit fixed, the depth cube or the 16-bit front-back format respectively.

We mainly made experiments with rigid objects, thus performing the depth map generation in a preprocessing step. Some tests have been made with deformable objects, forcing the depth map generation to be part of the simulation process. The generation of a complete depth map format with resolution 512^2 takes about 11, 17 and 26 ms using the 8-bit fixed, 16-bit front-back or the depth cube format respectively. Thus deformable objects should be applied only in combination with a small number of depth maps or particles.

Figure 6 visualizes the depth maps for two models: Two torii and the Stanford bunny. The torii are reconstructed using two depth cubes and two 16-bit front-back textures, giving 16 depth maps in total. The bunny is captured using one depth cube and one 16-bit front-back textures, giving 8 depth maps in total.

5.3. Hardware aspects

We use a standard normal index texture with resolution 256^2. Even though the L_1-parametrization would allow any application specific resolution, the handling of n bit integers or floats in the graphics hardware is hardly possible. Currently we use NVIDIA's pack/unpack functionality, which allows the packing of four bytes in one 32-bit float, for example. We would highly appreciate more functionality of this kind, e.g. to pack 5, 5, 6-bits in a 16-bit float.

Additionally, improved integer arithmetic and modulo operators would simplify the implementation of various shader functionality, e.g. the parallel sorting.

6. Conclusions and future work

A fully GPU based approach to realize the simulation and collision detection for large particle systems (PS) has been introduced. The simulation of PS is based on the "stream processing" paradigm, using textures to represent all data necessary to implement a state-preserving PS and collison detection using fragment shaders. The collision detection is based on depth maps. These are used to reconstruct an implicit model of the collider objects at the time of collision detection for particles. A novel technique to represent directional data was introduced and applied to store normal vectors using an indexing technique. When rendering the PS a parallel sorting algorithm can be applied to keep a proper particle order for non-commutative blending modes.

The proposed L_1-parametrization should be investigated further, especially its applicability to represent directional data, e.g. reflection maps. Additionally, investigations towards GPU based collision detection using polygons or more complex objects instead of particles should be made.

References

[ATI03] ATI TECHNOLOGIES INC.: *Normal map compression.* Tech. rep., ATI Technologies Inc., 2003. http://www.ati.com/developer/techpapers.html.

[Bat68] BATCHER K.: Sorting networks and their applications. In *Spring Joint Computer Conference, AFIPS Proceedings* (1968), pp. 307–314.

[BW03] BACIU G., WONG S.-K.: Image-based techniques in a hybrid collision detector. In *IEEE Trans. on Visualization and Computer Graphics* (2003), vol. 9, pp. 254–271.

[Dee95] DEERING M.: Geometry compression. In *ACM Proceedings SIGGRAPH* (1995), vol. 14, pp. 13–20.

[Gre03] GREEN S.: Stupid opengl shader tricks. http://developer.nvidia.com/docs/IO/8230/GDC2003_OpenGLShaderTricks.pdf, 2003.

[GRLM03] GOVINDARAJU N., REDON S., LIN M., MANOCHA D.: Cullide: interactive collision detection between complex models in large environments using graphics hardware. In *Proceedings of the ACM SIGGRAPH/EUROGRAPHICS Workshop on Graphics Hardware* (2003), Eurographics Association, pp. 25–32.

[Har03] HARRIS M.: *Real-Time Cloud Simulation and Rendering.* PhD thesis, Department of Computer Science, University of North Carolina at Chapel Hill, 2003.

[HS98] HEIDRICH W., SEIDEL H.-P.: View-independent environment maps. In *Proceedings of the ACM SIGGRAPH/EUROGRAPHICS Workshop on Graphics Hardware* (1998), ACM Press, pp. 39–45.

[HZLM02] HOFF K., ZAFERAKIS A., LIN M., MANOCHA D.: *Fast 3D Geometric Proximity Queries between Rigid and Deformable Models Using Graphics Hardware Acceleration.* Tech. Rep. TR-02-004, University of North Carolina at Chapel Hill, 2002.

[KJ01] KOLB A., JOHN L.: Volumetric model repair for virtual reality applications. In *EUROGRAPHICS Short Presentation* (2001), University of Manchester, pp. 249–256.

[KOLM02] KIM Y., OTADUY M., LIN M., MANOCHA D.: Fast penetration depth computation using rasterization hardware and hierarchical refinement. In *Proceedings ACM SIGGRAPH/Eurographics symposium on Computer animation* (2002), ACM Press, pp. 23–31.

[LG98] LIN M. C., GOTTSCHALK S.: Collision detection between geometric models: a survey. In *Proceedings of IMA Conference on Mathematics of Surfaces* (1998), pp. 37–56.

[McA00] MCALLISTER D.: *The Design of an API for Particle Systems.* Tech. rep., Dep. of Computer Science, University of North Carolina at Chapel Hill, 2000.

[NT99] NOORUDDIN F., TURK G.: *Simplification and repair of polygonal models using volumetric techniques.* Tech. Rep. GITGVU -99-37, Georgia Institute of Technology, Atlanta, 1999.

[NT03] NOORUDDIN F., TURK G.: Simplification and repair of polygonal models using volumetric techniques. *IEEE Trans. on Visualization and Computer Graphics9*, 2 (2003), 191–205.

[NVI03] NVIDIA CORPORATION: NVIDIA SDK. http://developer.nvidia.com, 2003.

[NVI04] NVIDIA CORPORATION: OpenGL extension EXT_pixel_buffer_object. http://oss.sgi.com/projects/ogl-sample/registry/EXT/pixel_buffer_object.txt, 2004.

[PBMH02] PURCELL T., BUCK I., MARK W. R., HANRAHAN P.: Ray tracing on programmable graphics hardware. In *ACM Proceedings SIGGRAPH* (2002), vol. 21, ACM Press, pp. 703–712.

[PDC*03] PURCELL T., DONNER C., CAMMARANO M., JENSEN H., HANRAHAN P.: Photon mapping on programmable graphics hardware. In *Proceedings of the ACM SIGGRAPH/EUROGRAPHICS Workshop on Graphics Hardware* (2003), Eurographics Association, pp. 41–50.

[Per03] PERCY J.: OpenGL Extensions. www.ati.com/ developer/techpapers.html, 2003.

[Ree83] REEVES W.: Particle systems - technique for modeling a class of fuzzy objects. In *ACM Proceedings SIGGRAPH* (1983), vol. 2, pp. 91–108.

[Sim90] SIMS K.: Particle animation and rendering using data parallel computation. In *Proceedings of the 17th annual conference on Computer graphics and interactive techniques* (1990), ACM Press, pp. 405–413.

[Ver67] VERLET L.: Computer experiments on classical fluids. i. thermodynamical properties of lennard-jones molecules. *Physical Review 159* (1967).

[VSC01] VASSILEV T., SPANLANG B., CHRYSANTHOU Y.: Fast cloth animation on walking avatars. In *Proc. EUROGRAPHICS* (2001), vol. 20, Eurographics Association, pp. 260–267.

Graphics Hardware (2004)
T. Akenine-Möller, M. McCool (Editors)

Understanding the Efficiency of GPU Algorithms for Matrix-Matrix Multiplication

K. Fatahalian, J. Sugerman, and P. Hanrahan †

Stanford University

Abstract

Utilizing graphics hardware for general purpose numerical computations has become a topic of considerable interest. The implementation of streaming algorithms, typified by highly parallel computations with little reuse of input data, has been widely explored on GPUs. We relax the streaming model's constraint on input reuse and perform an in-depth analysis of dense matrix-matrix multiplication, which reuses each element of input matrices $O(n)$ times. Its regular data access pattern and highly parallel computational requirements suggest matrix-matrix multiplication as an obvious candidate for efficient evaluation on GPUs but, surprisingly we find even near-optimal GPU implementations are pronouncedly less efficient than current cache-aware CPU approaches. We find the key cause of this inefficiency is that the GPU can fetch less data and yet execute more arithmetic operations per clock than the CPU when both are operating out of their closest caches. The lack of high bandwidth access to cached data will impair the performance of GPU implementations of any computation featuring significant input reuse.

Categories and Subject Descriptors (according to ACM CCS): I.3.1 [Computer Graphics]: Graphics processors

1. Introduction

The emergence of programmable graphics hardware has led to increasing interest in offloading numerically intensive computations to GPUs. The combination of high-bandwidth memories and hardware that performs floating-point arithmetic at significantly higher rates than conventional CPUs makes graphics processors attractive targets for highly parallel numerical workloads. Many applications beyond traditional graphics specific ones have been demonstrated to run on GPUs [HCSL02, BFGS03, KW03, MA03, BFH*04]. Particularly, algorithms that can be structured as *streaming* computations often realize notable performance gains. Streaming computations walk over a set of independent inputs and independently evaluate a kernel on each value. The combination of regular, predictable data access with the independence of each kernel invocation maps very nicely to graphics architectures.

Streaming computations can be characterized as being highly parallel and numerically intensive with little reuse of input data. However, many computations are both highly parallel and numerically intensive, but exhibit significant per element reuse. We study the efficiency of current graphics architectures in computations where each input contributes to many outputs. Optimized approaches to these problems on traditional CPUs are able to exploit cache hierarchies to effectively amplify memory bandwidth and keep processor functional units busy. We question whether calculations that benefit from memory "blocking" strategies will perform efficiently on GPUs and if analogous blocking strategies are necessary, or even possible, given the programming model and capabilities of current graphics architectures.

Specifically, we investigate dense matrix-matrix multiplication. It offers regular memory access and abundant parallel computation but features $O(n)$ data reuse and seems a natural candidate for a fast GPU implementation. Moreover, dense matrix-matrix multiplication is a building block of numerical libraries such as LAPACK [ABB*99]. These libraries are widely used in the development of scientific ap-

† {kayvonf, yoel, hanrahan}@graphics.stanford.edu

plications and must run efficiently if GPUs are to become a useful platform for numerical computing.

Cache-aware implementations for CPU architectures have been well studied [TCL98, WPD01] and several GPU algorithms have been developed. Larsen and McAllister [LM01] initially proposed an approach for computing matrix products on graphics hardware. They observed performance equaling that of CPUs but their computations were limited to 8-bit fixed-point data. Both Hall et al. [HCH03] and Moravánszky [Mor03] describe improved algorithms for modern hardware. Moravánszky reports his implementation is outperformed by optimized CPU code. Despite these results, little work has been done to analyze the performance limiting factors of these GPU implementations. We perform such an analysis and determine that the algorithms are bandwidth limited due to the inability of GPUs to provide high bandwidth access to frequently accessed data. Surprisingly, we find that existing algorithms are near-optimal for current hardware. That is, better performance will require architectural changes.

2. Implementations of Matrix-Matrix Multiplication

We consider the problem of computing the product, $C = AB$, of two large, dense, N×N matrices. We quickly describe naive and optimized CPU algorithms and then delve more deeply into solutions for a GPU.

2.1. Matrix-Matrix Multiplication on CPUs

The following CPU algorithm for multiplying matrices exactly mimics computing the product by hand:

```
for (i=0;i<N;i++)
   for (j=0;j<N;j++)
      C[i,j] = 0;
      for (k=0;k<N;k++)
         C[i,j] += A[i,k]*B[k,j];
```

Unfortunately, while simple, this algorithm suffers from poor locality. Elements of B are accessed columnwise, and therefore not in sequential order in memory (if matrices are stored in row-major order). While each iteration in j reuses row i of A, that row may have been evicted from the cache by the time the inner-most loop completes. As a result, the algorithm is bandwidth limited and displays poor performance and low efficiency (i.e. time spent doing computation versus loading data).

When sustaining cached accesses, the Pentium 4 can fetch a 128 bit SSE value (4 packed 32 bit floats) in a single cycle [HSU*01]. The basic flaw of the naive triple-for-loop implementation is that the size of its per-loop working set prevents bandwidth amplification from the closest (and fastest) levels of the memory hierarchy. A standard solution "blocks" the inputs into small submatrices that fit in the processor caches, and decomposes the computation of C into subcomputations on the submatrices. Additionally, B is often stored transposed to offer a more cache-friendly representation when walking its columns. The ATLAS [WPD01] software package, benchmarked in section 3, works even harder to achieve cache efficiency. ATLAS tests the host CPU to determine the sizes of system caches, then self-tunes its algorithms to use code snippets that are optimal for the detected cache sizes. Effective cache-aware implementations of matrix multiplication on CPUs achieve much higher effective bandwidth and hence numerical efficiency.

2.2. Matrix-Matrix Multiplication on the GPU

A simple approach to compute the product of two matrices on a GPU, although feasible only on architectures that support sufficiently long shaders, is to compute elements of the resulting matrix in a single rendering pass. We refer to this approach as **NV Single**. We store 2×2 blocks of matrix elements in 4 component texels as described by [HCH03]. This packing allows each input fetched from memory to be used twice in a fragment program. Thus the access of data used in the second operation is performed at register speed. Our shader, given below, reads 2 rows from matrix A and 2 columns of B to compute 4 elements of the output matrix. The values of variables i and j are provided to the shader via interpolated texture coordinates.

```
for (k=0;k<N/2;k++)
   C[i,j].xyzw += A[i,k].xxzz*B[k,j].xyxy +
                  A[i,k].yyww*B[k,j].zwzw;
```

Each shader program performs 4 inner loop iterations of the CPU algorithm given in section 2.1. GPU texture caches are organized to capture 2D locality, so sequential accesses along either axis are cache coherent. Additionally, shaders executing in parallel on adjacent fragments will fetch the same data from a row or column of the inputs simultaneously. In practice, for sufficiently large matrices, instruction counts exceed the limits of all architectures and a small number of passes are required to complete the computation.

In contrast, multipass algorithms are potentially more cache friendly with respect to inputs at the cost of performing extra framebuffer reads and writes. We implement a variation of the multipass algorithm presented by Larsen and McAllister [LM01]. Our implementation packs 4 consecutive elements from a matrix column into a texel, and thus performs a small number of 4×4 matrix by 4×1 vector products in a shader as described in detail by [Mor03]. This method makes 4x reuse of data in the texel obtained from matrix B, but only uses data fetched from A once. 6 fetches are needed to retrieve the operands needed for 4 `mad` operations, so in total this approach requires 1.5x more texture fetches than **NV Single**. However, since only a few rows and columns of the input textures are accessed in a given rendering pass, cache hit frequency is expected to be higher. A shader performing a single matrix-vector product is given given below.

It accepts i, j, and k arguments via interpolated texture coordinates. $accum[i, j]$ holds the partial result of $C[i, j]$ from the previous rendering pass. In each pass, k is incremented by 4.

```
C[i,j].xyzw = accum[i,j].xyzw +
              A[i,k].xyzw   * B[k/4,j].xxxx +
              A[i,k+1].xyzw * B[k/4,j].yyyy +
              A[i,k+2].xyzw * B[k/4,j].zzzz +
              A[i,k+3].xyzw * B[k/4,j].wwww;
```

Our NVIDIA implementation, **NV Multi**, computes a single matrix-vector product per shader. We unroll this kernel 3 to 6 times (the maximum possible amount allowed by the driver) to construct our shader for the ATI hardware, **ATI Multi**. We tested various amounts of unrolling on each platform. The chosen values maximized algorithm performance.

Performance tuning of algorithms on graphics hardware is difficult because vendors do not disclose specific architectural details, such as cache parameters or the physical layout of texture data in memory. We tested a variety of techniques intended to increase cache efficiency and leverage the parallel fragment processing architecture of the GPU. We varied the packing of matrix elements into textures, the unrolling of loops in shaders to reduce rendering passes, and utilized multiple render targets to perform multiplication of 4×4 submatrices of data within each shader. Additionally, we attempted to force blocked texture and framebuffer access across fragments by altering the size and order of geometric primitives sent to the card. We observed that the rasterization order of fragments produced by a single primitive covering the entire framebuffer is as efficient as any ordering created as a result of tiling the screen with geometry. Utilizing multiple shader outputs, despite allowing more computation to be performed per data access, yielded unsatisfactory results. Despite our optimization attempts, the two algorithms described above yielded the best performance.

3. Performance Results

We benchmarked our GPU algorithms and the CPU based matrix-matrix multiplication routine (sgemm) provided by ATLAS. Experiments were performed on a 3 GHz Pentium 4 CPU (512 KB L2 cache) featuring peak performance of 12 GFLOPS and an L1 cache bandwidth of 44.7GB/sec. We tested our GPU algorithms on the ATI Radeon 9800XT, a prerelease Radeon X800XT (500mhz core clock/500mhz mem), the NVIDIA GeForce FX 5900 Ultra, and a prerelease GeForce 6800 Ultra (350Mhz core/500Mhz mem), capable of achieving 26.1, 63.7, 40.0, and 43.9 GFLOPS respectively. We only measured the **NV Single** algorithm on the NVIDIA hardware because for large matrices its kernel requires more instructions than the ATI card supports.

We seek to specifically evaluate the computational performance of our GPU implementations and so explicitly exclude the overhead of repacking input matrices in system memory, transferring them to the graphics card, and initially

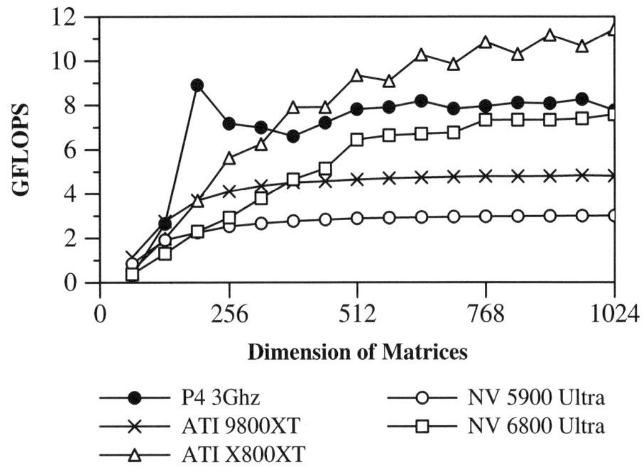

Figure 1: *Performance of multiplying square matrices on a 3 GHz Pentium 4, NVIDIA GeForceFX 5900 Ultra, prerelease GeForce 6800XT, ATI Radeon 9800XT, and prerelease Radeon X800XT.*

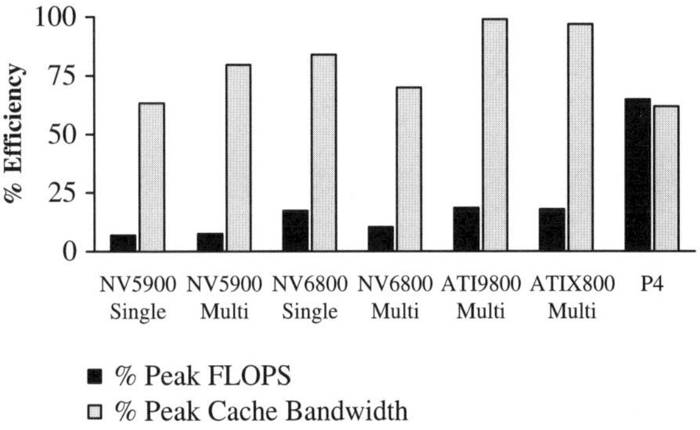

Figure 2: *Percentage computational and bandwidth efficiency when multiplying 1024x1024 matrices.*

loading and binding shader programs. In contrast, rather than attempt to instrument ATLAS code, we measure the entire time the ATLAS routine runs. ATLAS' initial self-tuning metacompilation, however, was performed during library installation and is not included in the timings. A summary of algorithm performance on the different platforms is given in Figure 1. GFLOPS values count each multiplication and each addition as distinct floating point operations (a `mad` instruction using `float4`s is thus eight operations).

ATLAS outperforms all GPU implementations except those running on the ATI X800XT despite being billed for any repacking or setup overheads and despite the significantly higher peak computational capability of all the graphics adapters. As shown in Figure 2, the ATLAS implementation achieves 65% efficiency while we observe no more than 17% efficiency on the NVIDIA chips and 19% on ATI platforms. Table 1 breaks down the multiplication of 1024×1024 matrices in more detail. In addition to the GFLOPS rate of each implementation, we computed effective input bandwidth. For the GPU implementations, we

	Time (s)	GFLOPS	BW (GB/s)
NV 5900 Single	0.781	2.75	7.22
NV 5900 Multi	0.713	3.01	9.07
NV 6800 Single	0.283	7.59	15.36
NV 6800 Multi	0.469	4.58	12.79
ATI 9800 Multi	0.445	4.83	12.06
ATI X800 Multi	0.188	11.4	27.5
P4 ATLAS	0.289	7.78	27.68

Table 1: *Measurements from the multiplication of 1024x1024 matrices.*

	GFLOPS	Cache BW	Seq BW
NV 5900	40.0	11.4	4.1
NV 6800	43.9	18.3	3.8
ATI 9800	26.1	12.2	7.3
ATI X800	63.7	28.4	15.6

Table 2: *Peak computation and bandwidth rates (cache hit and sequential fetch, in GB/sec) from our microbenchmarks.*

measure bandwidth as the total number of bytes fetched via `tex` instructions divided by total time to run our algorithm using a modified shader with nearly all math operations removed (drivers eliminate texture instructions if data is never used by shader, thus we cannot eliminate all math operations). Since we do not know the extent to which ATLAS is able to avoid memory accesses by reusing operands in registers, we compute its bandwidth as if all $2n^3$ accesses were from memory. In practice, we expect this significantly overestimates ATLAS' effective memory bandwidth. On the ATI platforms, bandwidth is computed assuming the transfer of 4 words per floating point value, although computation is performed at 24 bit precision.

Table 2 shows the results from a trio of microbenchmarks on our graphics cards. The first measures peak arithmetic performance by running a shader consisting entirely of `mad` instructions. The second measures peak bandwidth by repeatedly accessing texel (0, 0) from a number of input textures. We measure throughput approaching that of the theoretical peak bandwidth of the cards. The final microbenchmark runs a similar test, but walks the input textures sequentially by mapping each texel of the input textures to each rendered fragment. This is essentially the bandwidth realized by kernels that perform pure streaming over their inputs.

The key to the poor efficiency of the GPU algorithms can be seen by comparing Tables 1 and 2. The implementations are only able to utilize a small fraction of the GPUs' computational power despite data access patterns that yield high bandwidth. This inefficiency does not result from poor algorithm design or sub-optimal implementation. **ATI Multi** achieves over 95% of the bandwidth of perfect cache accesses yet cannot keep the computation units busy. The NVIDIA algorithms also leave chip arithmetic units idle despite accessing data at a large percentage of peak bandwidth. From the microbenchmarks and the cards' clock rates we compute that the 6800 Ultra and X800XT can perform 16 `mad`'s per clock and fetch 16 floats per clock. Since our four component `mad` takes 8 floats as input (and one register for the third operand), a shader with optimal memory access would require in excess of 8 math operations *per float fetched* in order to fully utilize the cards' computational capacity. Our matrix-matrix multiplication kernels cannot approach this ratio given current hardware constraints on instruction counts, register space, and the number of shader outputs.

4. Discussion and Conclusions

We expected that matrix-matrix multiplication would be a natural fit for the GPU and instead observe that optimized implementations achieve bandwidth approaching that of a theoretically optimal access pattern and yet achieve no better than 19% utilization of the arithmetic resources on our ATI hardware and no better than 17% utilization on our NVIDIA hardware. Our intuition proved completely misleading. Given current hardware designs, no algorithm will perform significantly better than existing approaches.

Indeed, we discover that available floating point bandwidth from the closest cache on a GPU is up to several times lower than that of current CPUs to their L1 caches. Without increasing this bandwidth (we expect the ratio of available computation to bandwidth to slowly worsen unless architectural design efforts specifically strive to reduce it), the efficiency of computing matrix-matrix products on graphics hardware will remain low.

One obvious way to increase performance is to perform more computation on data each time it is fetched by performing a level of blocking in the shader register file. The amount of potential reuse increases with the size of the submatrix multiplication performed in a shader. Current architectures provide sufficient registers to multiply two 6×6 matrices in a single shader, but do not permit shaders to generate more than a small number of outputs. As a result, we cannot use larger blocking within shaders to increase the compute to fetch ratio of our algorithms.

Multipass algorithms rely upon the ability to accumulate partial results into a final output. GPUs currently cannot accumulate into a full precision floating point framebuffer so our kernels must explicitly fetch the intermediate results produced by prior passes. However, this only accounts for 1/6 of the total input bandwidth of **NV Multi** and even less for **ATI Multi**. The benefit of adding hardware support for accumulation into floating point buffers pales in comparison to the advantages of significantly widening the path from the texture fetch units to the arithmetic units or the ability to do register level blocking.

We realize that when evaluating non-rendering algorithms on GPUs, it is important to remember that the design of graphics architectures is optimized for graphics workloads. Graphics applications primarily use 8-bit fixed point formats. Since four times as many values can be fetched per clock, compute to fetch ratios are much closer to those of conventional CPUs. Additionally, shaders often utilize interpolated values and constant parameters instead of texture memory for the majority of their inputs. Thus, graphics applications are more likely to perform many instructions per texture fetch. In contrast, practical numerical applications use full precision floating point values and depend upon iterating through input textures for access to data. Rebalancing GPUs to include wider or closer caches or an increased ability to operate on and output larger amounts of data per shader would make GPU architectures better suited for numerical workloads. Even streaming applications, already well suited for GPU architectures, would benefit from bandwidth improvements for regularly patterned texture accesses.

Although we investigate only matrix-matrix multiplication in detail in this paper, our arguments apply to many algorithms that feature data reuse. A wide variety of linear algebra operations rely upon blocking to amplify bandwidth and similar concerns apply to many general algorithms that repeatedly compute on large 2D (or n-D) pieces of memory. So long as graphics architectures continue to provide a low bandwidth connection between arithmetic units and nearby local memories, such algorithms will continue to suffer in efficiency when adapted to GPUs, despite increases in GPU clock rates and raw computational power, and despite the high levels of parallelism and numerical complexity inherent in the operations. Addressing these issues in future hardware designs will make GPUs a significantly more powerful platform for this broad class of applications.

5. Acknowledgments

We would like to thank ATI and NVIDIA for providing access to their latest hardware, and Sean Treichler of NVIDIA and Steve Morein for invaluable discussions about the design of graphics architectures. Support for this research was provided by the Rambus Stanford Graduate Fellowship and Stanford School of Engineering fellowship programs.

References

[ABB*99] ANDERSON E., BAI Z., BISCHOF C., BLACKFORD S., DEMMEL J., DONGARRA J., DU CROZ J., GREENBAUM A., HAMMARLING S., MCKENNEY A., SORENSEN D.: *LAPACK Users' Guide*, third ed. Society for Industrial and Applied Mathematics, Philadelphia, PA, 1999.

[BFGS03] BOLZ J., FARMER I., GRINSPUN E., SCHRODER P.: Sparse matrix solvers on the GPU: conjugate gradients and multigrid. *ACM Trans. Graph. 22*, 3 (2003), 917–924.

[BFH*04] BUCK I., FOLEY T., HORN D., SUGERMAN J., FATAHALIAN K., HOUSTON M., HANRAHAN P.: Brook for GPUs: Stream computing on graphics hardware. In *Proceedings of ACM SIGGRAPH 2004 (to appear)* (2004).

[HCH03] HALL J. D., CARR N. A., HART J. C.: Cache and bandwidth aware matrix multiplication on the GPU. *UIUC Technical Report UIUCDCS-R-2003-2328* (2003).

[HCSL02] HARRIS M. J., COOMBE G., SCHEUERMANN T., LASTRA A.: Physically-based visual simulation on graphics hardware. In *Proceedings of the ACM SIGGRAPH/EUROGRAPHICS conference on Graphics hardware* (2002), Eurographics Association, pp. 109–118.

[HSU*01] HINTON G., SAGER D., UPTON M., BOGGS D., CARMEAN D., KYKER A., ROUSSEL P.: The microarchitecture of the pentium 4 processor. *Intel Technology Journal 22*, 1 (2001).

[KW03] KRUGER J., WESTERMANN R.: Linear algebra operators for GPU implementation of numerical algorithms. *ACM Trans. Graph. 22*, 3 (2003), 908–916.

[LM01] LARSEN E. S., MCALLISTER D.: Fast matrix multiplies using graphics hardware. In *Proc. Supercomputing 2001* (2001).

[MA03] MORELAND K., ANGEL E.: The FFT on a GPU. In *Proceedings of the ACM SIGGRAPH/EUROGRAPHICS conference on Graphics hardware* (2003), Eurographics Association, pp. 112–119.

[Mor03] MORAVANSZKY A.: Dense matrix algebra on the GPU, 2003. http://www.shaderx2.com/shaderx.PDF.

[TCL98] THOTTETHODI M., CHATTERJEE S., LEBECK A. R.: Tuning strassen's matrix multiplication for memory efficiency. In *Proceedings of the 1998 ACM/IEEE conference on Supercomputing (CDROM)* (1998), IEEE Computer Society, pp. 1–14.

[WPD01] WHALEY R. C., PETITET A., DONGARRA J. J.: Automated empirical optimizations of software and the ATLAS project. *Parallel Computing 27*, 1–2 (2001), 3–35.

Committees and Reviewers

General Chairs

Michael Doggett
ATI

Anselmo Lastra
University of North Carolina

Program Chairs

Mark Harris
NVIDIA

David Luebke
University of Virginia

Papers Chairs

Tomas Akenine-Möller
Lund University

Michael McCool
University of Waterloo

Publicity Chair

Michael Meißner
Viatronix

Local Arrangements Chair

Sylvain Lefebvre
INRIA

Steering Committee

Wolfgang Straßer
University of Tübingen

Arie Kaufman
SUNY Stony Brook

Steve Molnar
NVIDIA

Paper Committee

Timo Aila
Helsinki University of Technology

Kurt Akeley
Stanford University and NVIDIA

Remi Arnaud
SONY

Paper Committee

David Baldwin
3Dlabs

Frank Crow
NVIDIA

Thomas Ertl
University of Stuttgart

Pat Hanrahan
Stanford University

John Hart
University of Illinois
Urbana-Champaign

Wolfgang Heidrich
University of British Columbia

Greg Humphreys
University of Virginia

Jan Kautz
MIT

Jim Kajiya
Microsoft

David Kirk
NVIDIA

Günter Knittel
University of Tübingen

Fons Kuijk
Epictoid

Tom Malzbender
Hewlett-Packard

Dinesh Manocha
University of North Carolina

Bill Mark
University of Texas

John Montrym
NVIDIA

Henry Moreton
NVIDIA

Paper Committee

Petri Nordlund
Bitboys

Marc Olano
University of Maryland

John Owens
University of California at Davis

Timour Paltashev
S3 Graphics

Ron Perry
MERL

Frans Peters
Philips Research

Andreas Schilling
University of Tübingen

Bengt-Olaf Schneider
NVIDIA

Peter Schröder
Caltech

Larry Seiler
ATI

Philipp Slusallek
Saarland University

Turner Whitted
Microsoft

Reviewers

Nate Carr
Dan Gelb
Michael Goss
Robin Green
Jesse Hall
Roy Hashimoto
Matthias Hopf
Zahid Hussain
Gundolf Kiefer
Antoine Labour
Ville Miettinen
Gabor Nagy
John Recker
Jörg Schmittler
Rüdiger Westermann

Cover Image Credits

Front Cover

Hardware-based Simulation and Collision Detection for Large Particle Systems
Andreas Kolb, Lutz Latta, Christof Rezk-Salama
pp. 123–132.

Back Cover

(top)

Squeeze: Numerical-Precision-Optimized Volume Rendering
Ingmar Bitter, Neophytos Neophytou, Klaus Mueller, Arie E. Kaufman
pp. 25–34.

(bottom left)

A Quadrilateral Rendering Primitive
Kai Hormann, Marco Tarini
pp. 7–14.

(bottom right)

UberFlow: A GPU-Based Particle Engine
Peter Kipfer, Marc Segal, Rüdiger Westermann
pp. 115–122.

Author Index

Color Plates

Analysis of required minimum bit precision for computation and storage for ray-casting and splatting.

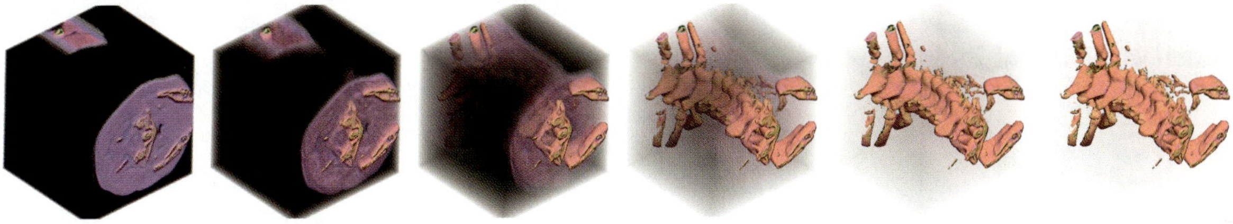

Ray casting 1024 samples per ray and compositing with $b = 8, 10, 12, 14, 15, 16$ bits precision (sample opacity=$\frac{1}{2^b}$).

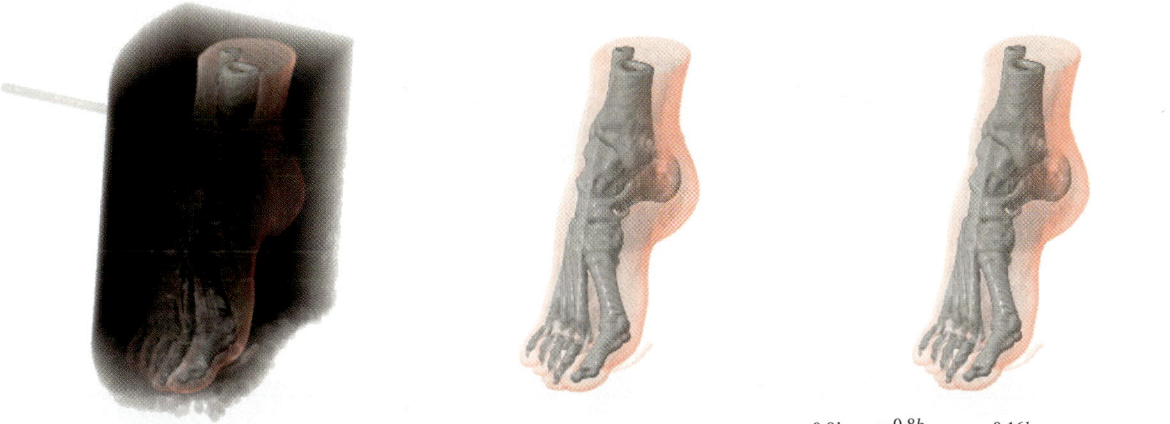

Splatting thin soft tissue using smallest representable non-zero opacity $\alpha^{0.8b}$, $\sqrt{\alpha}^{0.8b}$, and $\alpha^{0.16b}$.

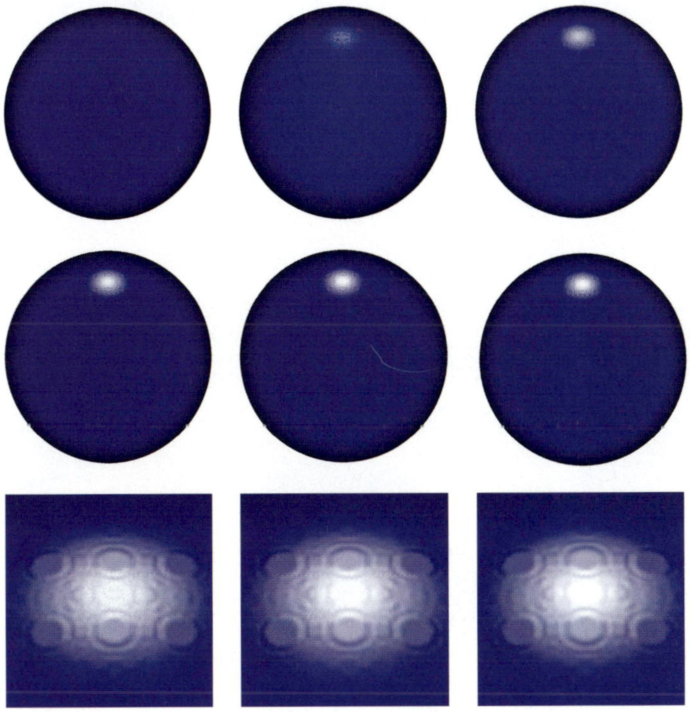

First two rows: 512^3 spheres with gradient precision of 4, 6, 8, 10, 12, and 14 bits. Third row: close-ups for gradient precision of 10, 12, and 14 bits.

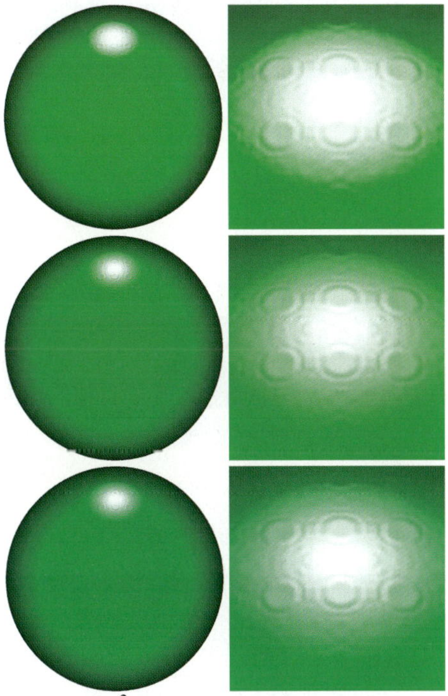

A 512^3 sphere with Knittel's Phong approximation (top), with float precision and the pow() function of Phong (middle), and with fixed-point Phong (bottom).

(a) Tile packing

(b) Result with only tile packing

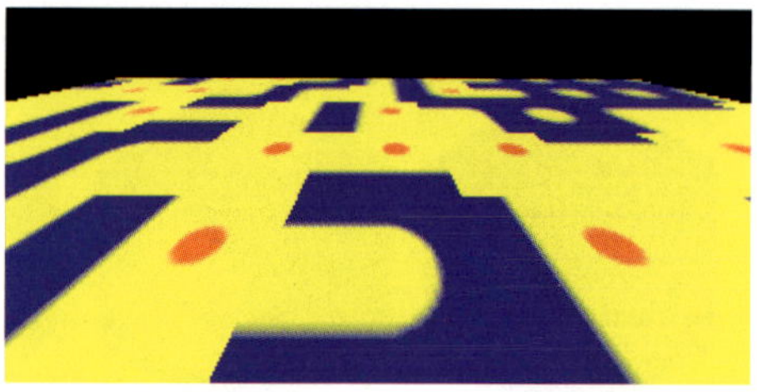
(c) Result with only corner packing

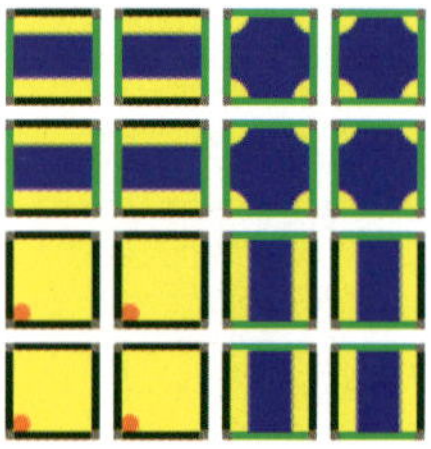
(d) Corner packing

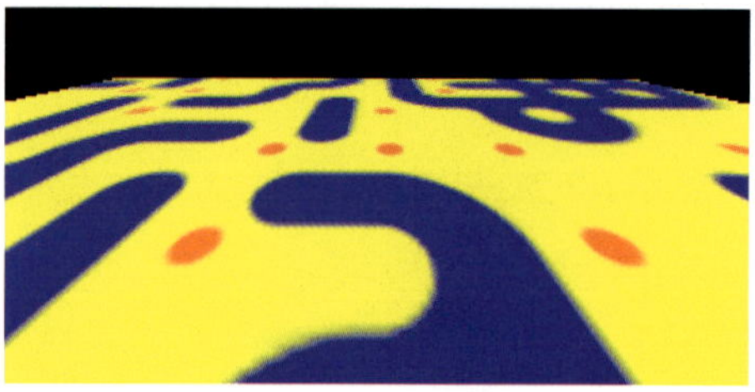
(e) Result with proper blending from (b) and (c)

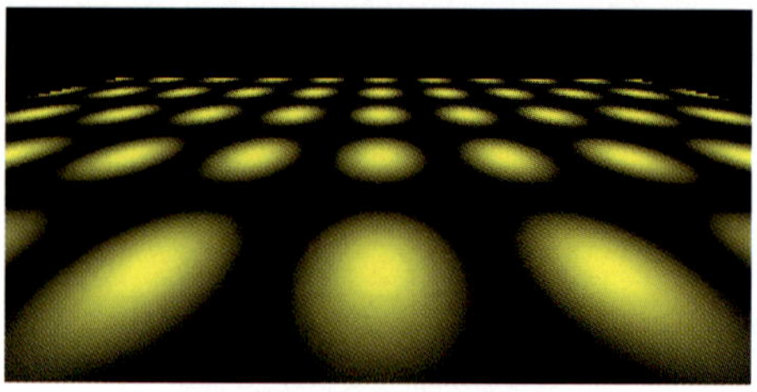
(f) Blending weight

Plate 1: Corner artifact example. The input tile set (a) contains 4 corner features shown as red dots. The results shown in (e) and (b) are rendering with and without corner handling, respectively. Note that result (b) contains noticeable errors around the upper right region of the polygon, while in result (e) the errors are much diminished. For reference, the result produced from only corner packing is shown in (c), and the blending weight is shown in (f).

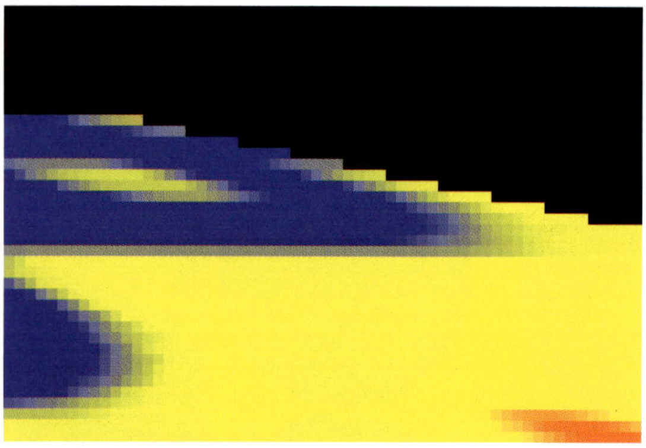
Detailed view of the upper-right region of (e)

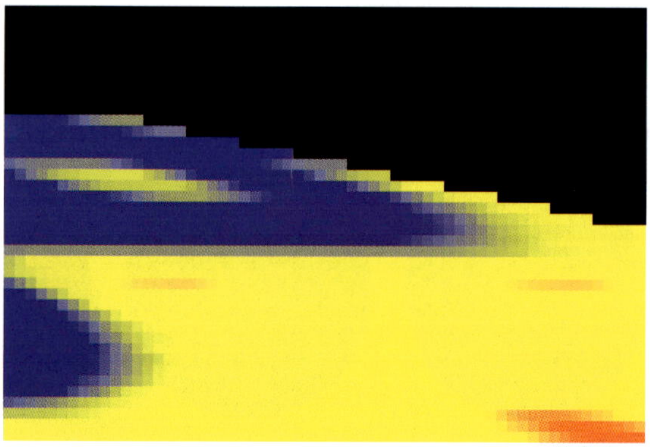
Detailed view of the upper-right region of (b)

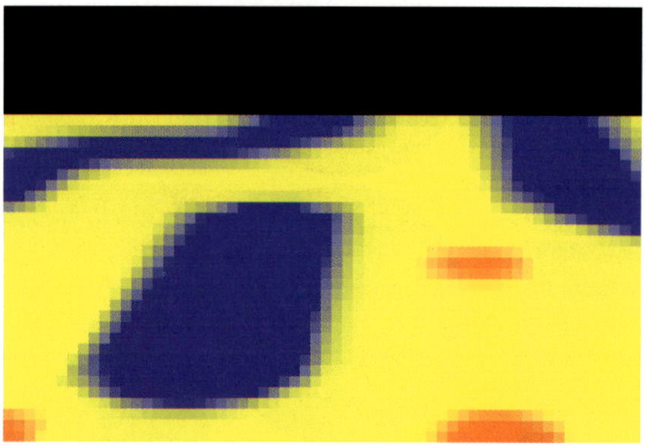
Detailed view of the upper-middle region of (e)

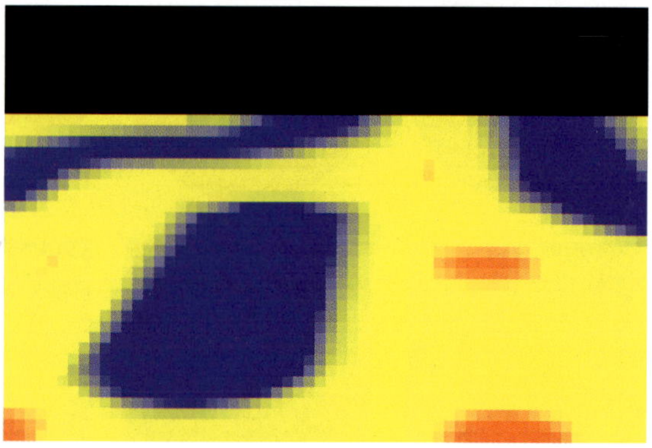
Detailed view of the upper-middle region of (b)

Plate 2: Detailed views of corner artifacts. Images on the right have more severe ghosting red lines, compared to images on the left.

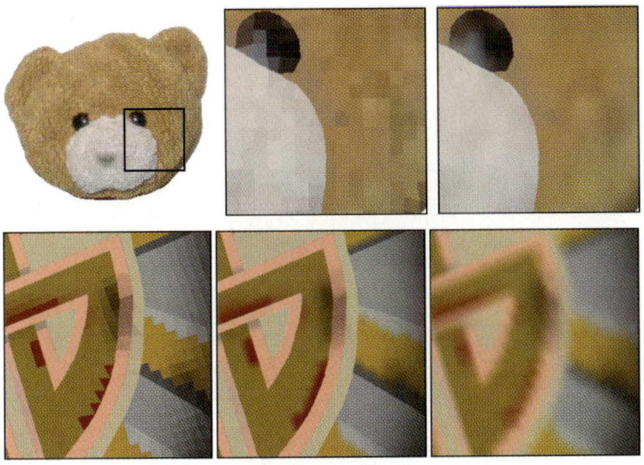

Figure 5: *Performing our algorithm on more complex textures highlights the need for proper filtering. In the top row, a photograph has been turned into a 64×64 silmap and the area indicated magnified to 512×512. The nearest neighbor silmap algorithm is shown on the left, the filtered version of the algorithm on the right. The bottom row shows unfiltered (left) and filtered (middle) closeups of the knight's shield. For reference, standard bilinear filtering disregarding discontinuities is shown on the right.*

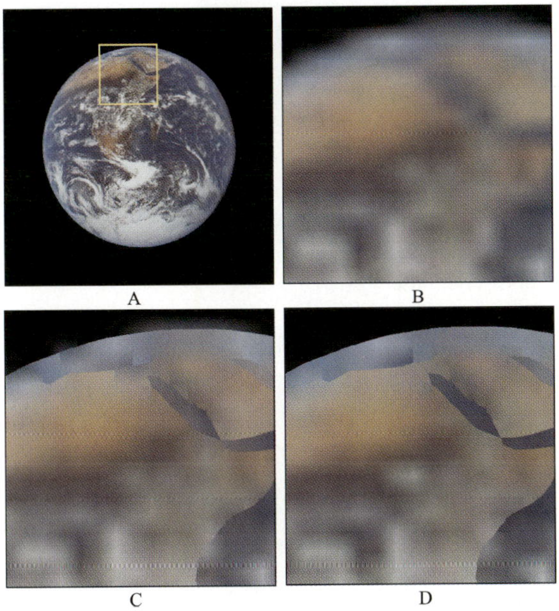

Figure 6: *Filtering a photograph to generate bitmap data for silmap texture. (A) Original photograph (1024×1024). (B) 64×64 pre-filtered bitmap magnified using standard bilinear interpolation. (C) Silmap technique, using a bitmap incorrectly pre-filtered ignoring discontinuities. (D) Results of our algorithm using a 64×64 silmap and properly filtering bitmap data. The improperly filtered image in (C) has a noticeable bluish tint in the space above the Earth and a brownish tint in the Red Sea.*

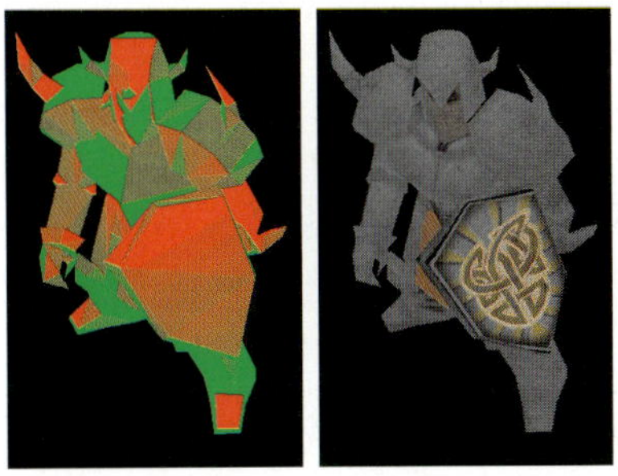

Figure 7: *(Left) The mipmap levels were rendered to switch to the silmap texture when necessary. The red regions indicate the areas that are being magnified. (Right) The result is a smooth blend of the silmap texture with mipmaps.*

Figure 8: *An example of a sign that might appear in a racing video game, with resolution 128×128. The parts shown are magnified to 512×512. (Left) Standard texturing with bilinear interpolation. (Right) Our technique.*

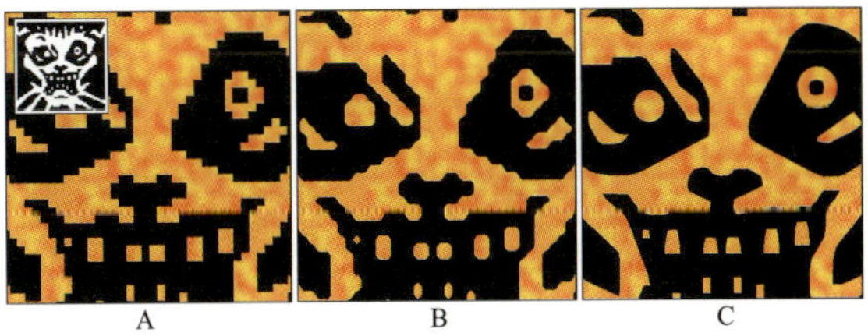

Figure 9: *The 64×64 bitmap shown in the inset is used to modulate a procedural noise shader. (A) Results of performing a standard, nearest-neighbor lookup on the texture. (B) In an effort to improve the quality, we interpolate the pixels and threshold the shader at 0.5, a technique often employed in games to improve the quality of textures used in this manner. (C) Using a silmap texture of the same resolution yields a visibly better result.*

Figure 6: *The evolution of a frame in the* Enemy Territories *game.*

Figure 4:
LEFT IMAGE: *Dense snow falling down on a landscape. Includes detection of ground contact.*
RIGHT IMAGES: *Effect of inter-particle collision (off at the top, on at the bottom).*

Figure 5: *Tracing large numbers of particles including particle-scene collision detection and front-to-back sorting to model natural phenomena. The point primitives can be rendered using any OpenGL functionality available.*

Figure 5: *Sample applications: "bunny in the snow" (left) and "Venus-fountain" (middle and right)*

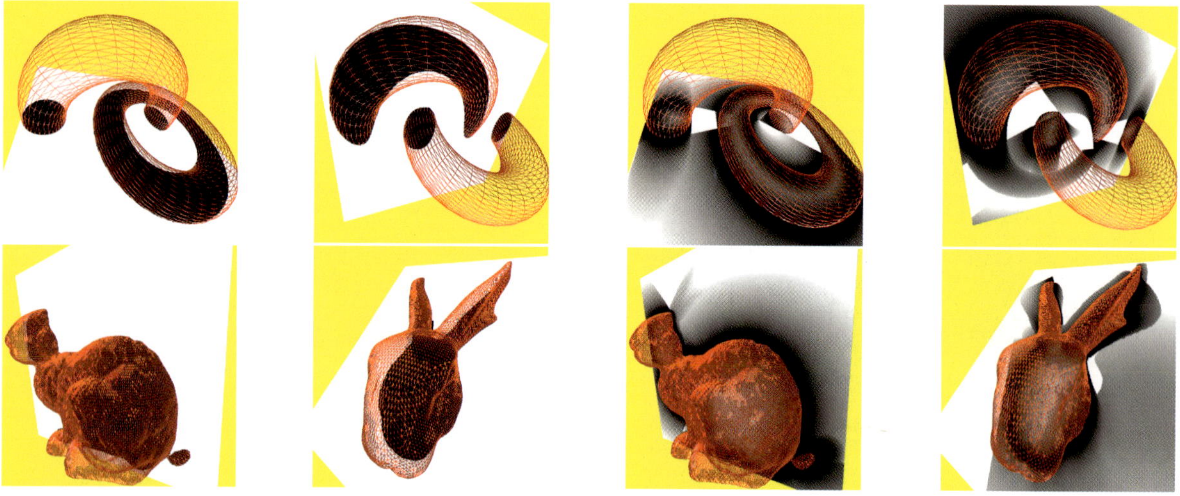

Figure 6: *Visualization of the implicit torii model (top row) and the implicit bunny model (bottom row) along slices: Interior/exterior classification (left) and approximate distance map plotting absolute distance values (right). Additionally, the wireframe model is rendered.*